American Institute of Chemical Engineering

Energy Efficiency for a Sustainable Economy 2005

Held at the 2005 AIChE Spring National Meeting

April 10-14, 2005
Atlanta, Georgia, USA

Printed from e-media with permission by:

Curran Associates, Inc.
57 Morehouse Lane
Red Hook, NY 12571
www.proceedings.com

ISBN: 978-1-60423-467-1

Some format issues inherent in the e-media version may also appear in this print version.

TJ - Energy Efficiency for a Sustainable Economy

Session 117 - Tutorial - Energy-Water Nexus in the Process Industry
Chair: C B. Panchal
Vice Chair: Peter R. Pujado

1 **Energy-Water Nexus and the Process Industry**
John R. Gasper

Session 118 - Process Integration for Energy Conservation
Chair: Ronald W. Breault
Vice Chair: V K. Mathur

2 **Process Integration for Energy Conservation**
Paul Tripathi

Session 119 - Mass and Heat Transport in Fluidized Bed Energy Systems
Chair: Ronald W. Breault
Vice Chair: Thomas Ho

3 **A Mass Transfer-Based Kinetic Model for Simulating a Microwave-Heated Fluidized Bed Mercury Desorber**
Anand Makhija, Jack R. Hopper, Jerry Lin, Tae-Hoon Kim and Thomas Ho

4 **Sensitivity of Gas-Solids Dispersion and Mass Transfer Coefficient in an Eulerian-Eulerian CFD Model**
Chris P. Guenther and Ronald W. Breault

17 **Mass Transfer in Circulating Fluidized Beds**
Rex Compston and Ronald W. Breault

29 **Enhanced Productivity of Chemical Processes Using Dense Fluidized Beds**
Paul R. Tortora, Timothy J. O'Hern, Ken Williams and Dale Snider

Session 120 - Renewable Sources of Energy I
Chair: V K. Mathur
Vice Chair: Gerald R. Nix

40 **Numerical Investigation of Water and Thermal Management in PEM Fuel Cells**
Jinquan Xu and M. Yousuff Hussaini

41 **Carbon Textiles and Their Use in Fuel Cell Gas Diffusion Layers**
Jim Crawford, Paul Kennedy

75 **Role of Carbon in a PEM Fuel Cell System**
Jim Crawford, V. K. Mathur and Xiaofeng Xie

79 **The Role of Thermodynamics and Rate Processes with Respect to Alternative Fuels**
Michael Neuman and Stuart W. Churchill

80 **Hydrogen Production by Dissociation of Water through a Novel Membrane**
Jack A. Shindle and Reinder J. Boersma

Session 121 - Microchannel Transport Phenomena - Fundamentals and Applications
Chair: Masahiro Kawaji
Vice Chair: Mohamed S. El-Genk

92 **On a Explicit Relation for Plasma Layer Thickness**
Kal Renganathan Sharma

97 **The Influence of Bend Geometry on the Performance of a MicroGC**
Adarsh D. Radadia, Keith R. Cadwallader, Mark A. Shannon and Rich Masel

98 **The Effects of Inlet Geometry on Gas-Liquid Two-phase Flow in Microchannels**
Koji Mori and Masahiro Kawaji, Daniel Bolintineanu and Y. Shibata

107 **Fabrication and Performance of Pulsating Micro Heat Pipes**
Christian G. Lu, Kathryn Nikkanen and Masahiro Kawaji

115 **Numerical Study on the Gas and Liquid Slugs for Taylor Flow in a Microchannel**
Adeniyi Lawal and Dongying Qian

Session 122 - Renewable Sources of Energy II
Chair: Gerald R. Nix
Vice Chair: V K. Mathur

116 **Conversion Optimization for Geothermal Power Plants**
Gerald R. Nix

117 **Pyrolysis of Alfalfa Stems and Dedicated Herbaceous Energy Crops in the National Biomass Initiative Program**
Akwasi A. Boateng

118 **Concentrated Acid Pretreatment for the Conversion of Lignocellulosic Materials to Sugar**
Roger D. Hester and William S. Miller

119 **Feedstock Flexible Fluidizable Reforming Catalyst Development**
Esteban Chornet, Kimberly A. Magrini, Richard French, Stefan Czernik and Yves Parent

120 **Solar-thermal Decarbonization of Natural Gas – A Near Term Process**
Alan Weimer, Allan Lewandowski and Raymond Hobbs

Session 123 - Energy Tools and Methodologies for Industrial Applications
Chair: Frank Zhu
Vice Chair: Kirtan K. Trivedi

121 **New Opportunities for Thermal Energy Saving in BPA Chemical Plants Using Real-Time Energy Integration**
Abd ElRazak El-Saleh, Kamrul Hasan and Mahmoud Bahy M. Noureldin

122 **Energy Optimization Tools for Non-core Business Applications**
Anthony L. Wright, Frederick P. Fendt and Glenn Hahn

131 **Innovative Absorber/Stripper Configurations for CO2 Capture by Aqueous Monoethanolamine**
Gary T. Rochelle and Majeed S. Jassim

132 **Automated Design and Optimisation of Flexible Utility Systems under Variable Operating Conditions**
Oscar Aguilar and Robin Smith, Simon Perry, and Jin-Kuk Kim

141 **Systematic Driver and Power Plant Selection for Power-demanding Industrial Processes**
Frank L. Del Nogal, Jin-Kuk Kim, Robin Smith and Simon J. Perry

156 **Model Centric Approach for Energy Management and Optimization**
Jose Desouza, Nii Asante and Piyush Shah

157 **Future Trends and Practical Solutions for Industrial Energy Optimization**
Frank Zhu

158 **Determining Your Plant Energy Profile Using the PEP Software Tool**
James Quinn

Session 124 - Energy Efficiency for Natural Gas Utilization
Chair: Lin Li
Vice Chair: Peter R. Pujado

166 **BP/DPT's Efficient Gas Conversion Route**
Joep J H M Font Freide, and Tim Gamlin

167 **Overcoming the Challenges of GTL commercialisation**
Theo Pretorius

196 **Advanced Energy Integration in Reforming Technology or Taking the Power Station out of Steam Reforming**
John Swinney

202 **Large Scale Production of Synthesis Gas for Methanol and GTL Plants**
Kim Aasberg-Petersen, Martin R. Ostberg, Per K. Bakkerud

203 **Steam Cracking and Natural Gas to Olefins: Energy Efficiency and CO2 Emissions**
Tao Ren, *Martin Patel and Kornelis Blok*

Session 125 - Improvements in Energy Efficiency in Process Plants
Chair: Paul Tripathi
Vice Chair: Phil Lewellen

237 **Simplified Heat Transfer Equations for Rotary Kilns**
Louis E. Herrington

238 **Effective Plant-wide Energy Management and Optimization**
Annelize Victor

251 **Techno-Economic Evaluation of an Improved and Energy Efficient Natural Gas Liquid (NGL) Removal Process**
Aqil Jamal, Dennis Leppin, Howard Meyer, Javad Abbasian and Yingzhong Lu

252 **Save Energy Without Capital Expenditure First before investing**
Paul Tripathi

253 **Improved Energy Efficiency Using Regenerative Oxidizers for VOC Control**
Michael J. Hager

254 **Energy Saving Opportunities in Heat-integrated Crude Oil Distillation Systems**
Megan Jobson, Robin Smith and Vikas Rastogi

256 **Development of Advanced Membranes Technology Platform for Hydrocarbon Separations**
Dilip Kalthod

257 **Ultra-Low Excess Air Control for Process Heaters**
Kurt Jasin

117a Energy-Water Nexus and the Process Industry

John R Gasper

Competition for water threatens current and future energy supplies. Growing amounts of energy are required to meet water treatment and pumping needs. Conflicts at this energy-water interface or nexus can limit the competitiveness of U.S. industry. The ability to resolve these conflicts can be enhanced through R&D in assessment, basic science and technology. Congress and the federal government are developing R&D programs that focus on the energy-water nexus. The process industry will be among the sectors of the U.S. economy that can play a role in these R&D programs. This tutorial will explore the relationship between energy and water and the related R&D planning.

118a Process Integration for Energy Conservation

Paul Tripathi

Process integration is an important tool to conserve energy. These days of high energy prices and global competition the process industry is looking for ways to reduce cost. Process integration provides a systematic approach to reduce energy consumption. The additional benefit derived from the energy conservation is the reduction in greenhouse gases.

The topics in this tutorial are:

-Laws of thermodynamics

- Exergy Analysis

- Pinch Analysis

-Concept of Marginal Cost of Utility

-Examples of Process Integration using Pinch Analysis

-Practical limitations of the concept

119a A Mass Transfer-Based Kinetic Model for Simulating a Microwave-Heated Fluidized Bed Mercury Desorber

Anand Makhija, Jack R Hopper, Jerry Lin, Tae-Hoon Kim, Thomas Ho

Control of mercury emissions from various combustion sources has attracted great attention due to the toxic nature of mercury and the current and potential regulations. The most effective method for controlling trace elemental mercury emission is to employ activated carbon (AC) to adsorb mercury from the combustion flue gas. However, an environmental concern is the production of large quantity of spent activated carbon contaminated with various pollutants. A practical solution is to regenerate the AC for reuse in a conventionally- or microwave-heated fluidized bed mercury desorber. In this study, a mass transfer-based mercury desorption model was employed to simulate the performance of a microwave-heated fluidized bed mercury desorber. The desorber had a 0.3 m ID and was filled with various types of activated carbon (AC) contaminated with various concentrations of mercury. Nitrogen was used as the desorption gas and the simulation parameters included microwave power level, fluidization velocity, AC type, AC size, AC amount, and concentration of mercury contamination. The objective of the simulations were to identify optimal operating conditions resulting in the fastest desorption rate as well as the highest mercury concentration in the effluent gas. The latter is essential for effective mercury recovery from the effluent stream. The detailed results will be presented at the meeting.

Sensitivity of Gas-Solids Dispersion and Mass Transfer Coefficient in an Eulerian-Eulerian CFD Model

Chris Guenther and Ron Breault**
** National Energy Technology Laboratory*
Morgantown, WV, 26507

Abstract

Eulerian-Eulerian CFD models have become widely accepted in the modeling of dense gas-solids systems. These models have been used to predicted bubble dynamics in fluidized beds and clustering and core annular structure in circulating fluidized beds. However, coupling the hydrodynamics predicted by these models with chemical reactions is relatively new application. In a recent review paper by Breault 2004 gas-solids dispersion coefficient reported in the literature differ by 5 orders of magnitude and mass transfer coefficients differ by 7 orders of magnitude. Given such variability in these coefficients this paper investigates the sensitivity of these values in the Eulerian-Eulerian model MFIX (Multiphase Flow with Interphase eXchanges; www.mfix.org). Two dimensional simulations based on a circulating fluidized bed (CFB) in operation at the National Energy Technology Laboratory (NETL) were conducted. The simulations were conducted in the core-annular regime with carbon combustion as the single chemical reaction. The sensitivity in the predicted carbon conversion as a function of gas-solids dispersion and mass transfer coefficient is reported.

Introduction

Circulating fluidized bed technology (CFB) is being used in a variety of metallurgical, chemical and energy processes. In the energy sector, one of the promising applications of CFB's is the use of transport gasifiers in advanced coal-fueled power systems. Power plants of the future e.g., Clean Coal Power Initiative (CCPI) and FutureGen power plants will both use CFB technology to design, scale-up and operate transport gasifiers. Transport gasifiers because of their ability to operate at high throughput conditions, achieve better mixing, increase mass and heat transfer, and achieve high carbon conversion will be a key component in the overall plant design. Given such an important role these gasifiers will have on energy plants of the future, understanding the hydrodynamics and the coupling between the hydrodynamics and chemistry is critical for their design, scale-up, and operation.

To understand the hydrodynamics in a transport gasifier heavily instrumented cold flow CFB's are routinely used on a variety of scales. Experimental and pilot scale hot units are also used to understand the chemistry and heat transfer and its dependency on the hydrodynamics. This information is obviously important, but there is no well established method for using this information for scale-up to a commercial design. Computational fluid dynamics (CFD) and Eulerian-Eulerian two-fluid models are beginning to show promise as a tool to aid engineers in the design, scale-up and operation of CFB's. There are a number of papers in the literature validating these models against cold flow units (Guenther et al. 2002, Gidaspow et al. 1992, and Tsuo and Gidaspow 1990) to name just a few. On the other hand, very little has been reported in the literature using Eulerian-Eulerian two-fluid models to simulate gas-solids systems with chemical reactions. Guenther et al. 2001 looked at the numerical effects on two and three-dimensional simulations of silane pyrolysis in a fluidized bed. Samuelbsberg 1994 considered simulations of fluidized bed reactors in his Ph.D. thesis. Recently, Guenther et al. 2003a and Guenther et al. 2003b used the two-fluid model MFIX to simulate a large pilot scale transport gasifier in operation at Power Systems Development Facility (PSDF) near Wilsonville, Alabama. This work is a joint project of the U.S. Department of Energy Southern Company Services, Inc., and Kellogg Brown & Root. These transient three-dimensional simulations considered both bituminous (Hiawatha) and sub-bituminous (PRB) coals under both air and oxygen blown conditions.

In the previous investigations done by the author simulating the transport gasifier at PSDF no effort was given to the sensitivity of the model's results on the dispersion and mass transfer models in MFIX. This paper motivated the authors, by the review done by Breault 2004, to investigate the models sensitivity in the predicted carbon conversion as a function of gas-solids dispersion and mass transfer coefficient. Two-dimensional simulations of a circulating fluidized bed operating in the core-annular regime with carbon combustion as the single chemical reaction was considered. The goals of this investigation were to first understand which coefficients had a dominant effect on carbon conversion and then to run a sufficient number of simulations to correlate the predictive carbon conversion with a range of coefficient values.

Computational Model

The MFIX model has been in use at the National Energy Technology Laboratory (NETL) for over fifteen years and has been become internationally recognized as one of the premier two-fluid models available to researchers. Its open source format and FORTRAN coding of subroutines and versatile post-processing tools makes MFIX an ideal platform to develop, validate and test sub-models (e.g., dispersion and mass transfer coefficients) within a two-fluid framework. Two-fluid hydrodynamic models also referred to as Eulerian-Eulerian models, treat the fluid and solids as two continuous and fully miscible phases. This approach results in mass, momentum, and energy balance equations for both the gas and solids phases. For isothermal conditions, the continuity, momentum, and species balance equations are given below.

Gas-Phase Continuity

$$\frac{\partial}{\partial t}(\varepsilon_g \rho_g) + \nabla \circ (\varepsilon_g \rho_g \vec{v}_g) = \sum_{n=1}^{N_g} R_{gn}$$

Solids-Phase Continuity

$$\frac{\partial}{\partial t}(\varepsilon_s \rho_s) + \nabla \circ (\varepsilon_s \rho_s \vec{v}_s) = \sum_{n=1}^{N_s} R_{sn}$$

Gas-Phase Momentum

$$\frac{\partial}{\partial t}(\varepsilon_g \rho_g \vec{v}_g) + \nabla \circ (\varepsilon_g \rho_g \vec{v}_g \vec{v}_g) = -\varepsilon_g \nabla P_g + \nabla \circ \overline{\overline{\tau}}_g + F_{gs}(\vec{v}_s - \vec{v}_g) + \varepsilon_g \rho_g \vec{g} - R_0(\xi_0 \vec{v}_g + \overline{\xi_0} \vec{v}_s)$$

Solids-Phase Momentum

$$\frac{\partial}{\partial t}(\varepsilon_s \rho_s \vec{v}_s) + \nabla \circ (\varepsilon_s \rho_s \vec{v}_s \vec{v}_s) = -\varepsilon_s \nabla P_g + \nabla \circ \overline{\overline{S}}_m - F_{gs}(\vec{v}_s - \vec{v}_g) + \varepsilon_s \rho_s \vec{g} - R_0(\xi_0 \vec{v}_g + \overline{\xi_0} \vec{v}_s)$$

Species Balance

$$\frac{\partial}{\partial t}(\varepsilon_m \rho_m X_{mn}) + \nabla \circ (\varepsilon_m \rho_m X_{mn} \vec{v}_m) = R_{mn}$$

where $m = g$ or s for the gas or solids phase and $\overline{\xi_0} = 1 - \xi_0$ and $\xi_0 = 1$ if $R_0 < 0$; else $\xi_0 = 0$. The eight dependent hydrodynamic variables in 3D: void fraction ε_g (the solids fraction $\varepsilon_s = 1 - \varepsilon_g$), pressure P_g, and six velocity components are found by using MFIX to numerically solve the coupled non-linear partial differential equations. The number of species mass fractions (X_{mn}) tracked are given in the chemistry model. Constitutive relations needed to close the system, and the gas/solids energy balance equations can be found in Syamlal et al. 1993 and Syamlal 1998. A discussion on the solution procedure and further numerical references can be found in Guenther and Syamlal 2001.

Chemistry Model

In this investigation to isolate the effects of the gas/solids dispersion and mass transfer coefficients on carbon conversion the only reaction considered in the simulations is carbon combustion given by the heterogeneous reaction $2C + O_2 \rightarrow 2CO$. The reaction rate is given by

$$\frac{-6\varepsilon_s P_{O_2}}{d_p\left(\frac{1}{k_f} + \frac{1}{k_a} + \frac{1}{k_r}\right)} (mol/cm^3 s$$

where the film resistance is given by

$$k_f = \frac{D_{O_2} Sh}{d_p R_{O_2} T_f}$$

the Sherwood number is given by Gunn 1978

$$Sh = (7 - 10\varepsilon_g + 5\varepsilon_g^2)(1 = .7\,\mathrm{Re}^{0.2}\, Sc^{1/3}) + (1.33 - 2.4\varepsilon_g + 1.2\varepsilon_g^2)\mathrm{Re}^{0.7}\, Sc^{1/3}$$

and the Schmidt number is given by

$$Sc = \frac{\mu_g}{\rho_g D_{O_2}}.$$

The ash layer resistance is given by

$$k_a = \frac{2r_d D_e}{(1 - r_d) d_p R_{O_2} T_s}$$

and the ratio of core diameter to particle diameter is

$$r_d = \left(\frac{X^0_{s,ash} X_{s,carbon}}{X^0_{s,carbon} X_{x,ash}} \right)^{1/3}.$$

The surface reaction is given by Desai and Wen 1978

$$k_r = 8710 \exp\left(-\frac{27000}{1.987T_s} \right) r_d^2.$$

Simulation Results and Discussion

Based on the flow conditions in a large scale cold flow CFB in operation at the NETL, the initial investigation considered two-dimensional simulations of a conceptual design riser section of a CFB. The gas and solids flow rate was adjusted to produce a core-annular flow regime similar to the flow regimes found in the CFB at NETL. Core-annular flow is characterized by growth and destruction of groups of particles called clusters (Breault et al. 2005). These clusters tend to form near the walls of the riser and due to their mass can no longer be supported by the upward flow of gas fall down along the wall. Eventually these clusters break apart as they travel down and are entrained back into the upward gas flow. This constant formation and destruction of clusters produces a core-annular flow regime where near the wall (annulus) high solids concentrations and downward gas/solids velocities are found and in the center of the riser (core) low solids concentrations and high upward gas/solids velocities are found. The flow conditions are given in Table 1 and the schematic of the simulation geometry is given in Figure 1.

	Gas and Ash In	Gas and Coal In
ρ_s, d_p (g/cm^3, microns)	0.190, 0.9	0.190, 0.9
ε_g	0.6	0.9
$\vec{v}_g$ (cm/s)	175.0	-1.0
X_{ash}, X_{carbon} (mass fraction)	1.0, 0.0	0.6, 0.4
T_g, T_s (Kelvin)	1144, 1144	294, 294

Table 1 Two-Dimensional Simulation Flow Boundary Conditions

Initially the riser section was void of any solids, filled with air and at 1144K. The simulations kept the boundary conditions fixed throughout the simulation. The transient simulations were run for 70 seconds solving for the voidage, gas/solids velocity components, energy, and species mass fraction and considered only a single particle size for both the hydrodynamics and chemistry. Results were taken at the exit time averaging over the last 40 seconds of the transient data.

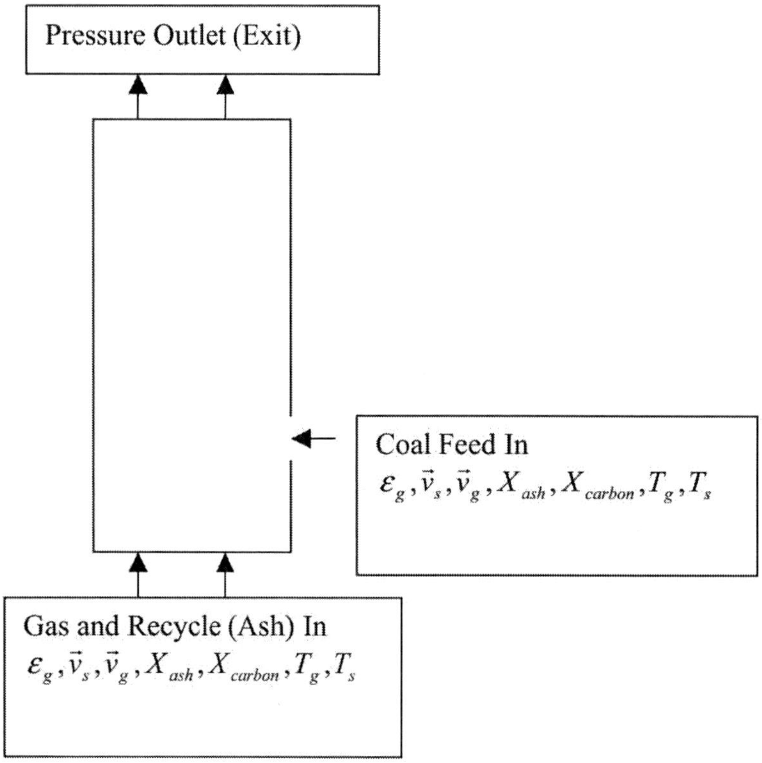

Figure 1 Two-dimensional simulation geometry

To investigate the values of mass transfer and gas/solids dispersion coefficients two sets of simulations were conducted using the low (0.0005) and high (10.0) values found in the literature for mass transfer. These constants were then hard-wired into the MFIX code replacing the default model for mass transfer i.e., Gunn 1978. Then for the low and high mass transfer numbers the gas and solids dispersion coefficients were varied to reflect the values found in the literature. Table 2 summarized the high and low values for the gas/solids dispersion coefficients found in the literature along with the acronyms used in this paper.

		Dispersion Coefficients			
		High Dispersion		Low Dispersion	
		High Gas HG	High Solids HS	Low Gas LG	Low Solids LS
Gas Phase	Axial Gas	5.0		0.05	
	Radial Gas	0.05		0.0005	
Solids Phase	Axial Solids		5.0		1.0
	Radial Solids		0.5		0.08

Table 2 Low and High Values for the gas/solids Dispersion Coefficients.

Figure 2. shows the carbon conversion results for the low Sherwood number and Figure 3 shows the same results with a high Sherwood number. The vertical axis represents the percentage of carbon conversion based on the time averaged mass flow rate of carbon in the gas phase leaving the reactor.

$$carbon\ conversion = \left(1 - \frac{\dot{M}_C^{in} - \dot{M}_C^{out}}{\dot{M}_C^{in}}\right) \times 100\%$$

where $\dot{M}_C^{in}$ and $\dot{M}_C^{out}$ are the mass flow rate of the carbon in the solids phase entering the riser (fixed boundary condition) and mass flow rate of the carbon in the gas phase exiting the reactor (time averaged).

Comparing the results presented in Figure 2 and in Figure 3, it can be seen that regardless of the value of the Sherwood number the effect of changes to the dispersion coefficients have the same trend-wise effect on the predicted carbon conversion. Comparing the cases with high gas dispersion to low gas dispersion, reveals that lower values of the gas dispersion coefficient have a significant impact on the conversion with the low value predicting carbon conversions 6 to 8 times that of the high gas dispersion coefficient. The opposite trend is observed for changes in the solids dispersion coefficient; however the magnitude of the change to the predicted carbon conversion is significantly less, being only on the order of 20 to 30 percent.

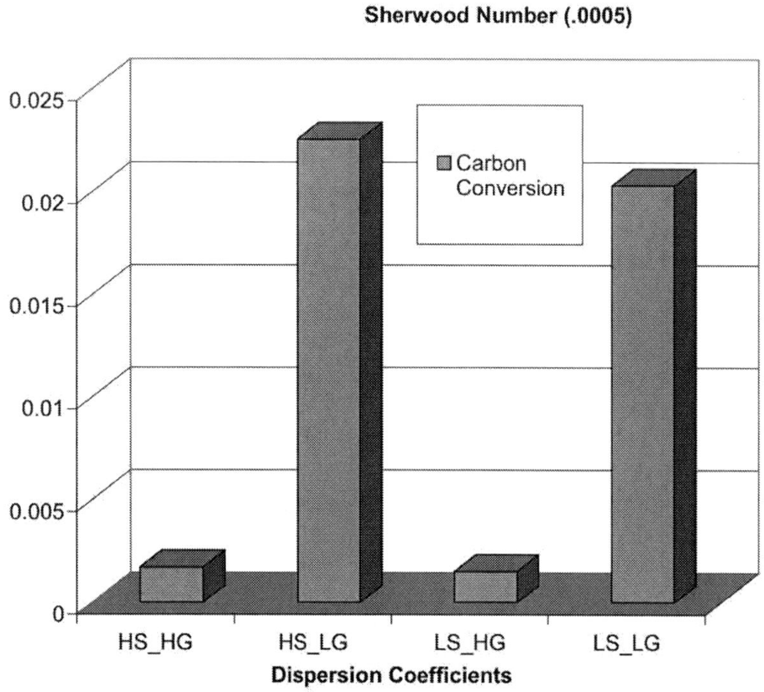

Figure 2 Carbon conversion with a low Sherwood number.

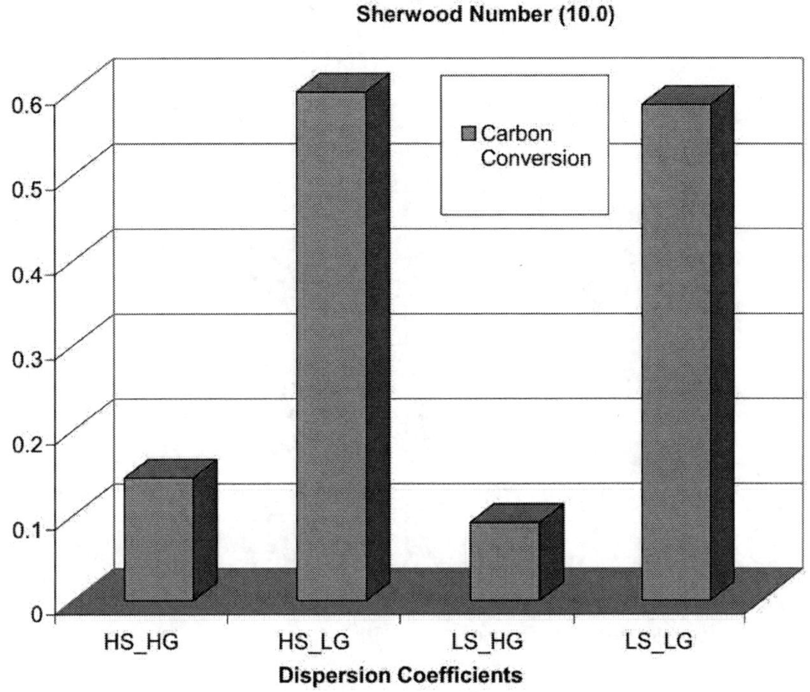

Figure 3 Carbon conversion with a high Sherwood number.

Figure 4 compares the data for high solids dispersion and high Sherwood number for several values of the gas dispersion coefficient. These being the "High Value", the average literature value from the Breault (2004) review, the "Low Value", and the nominal value from the MFIX default correlation. As can be seen, the carbon conversion for the average literature value falls in between the values for the "High" and "Low" gas dispersion coefficients. The carbon conversion for the prediction from the MFIX default correlation is higher than that for the "Low" case. Based upon these results, there is a strong likelihood that MFIX, for these kinetics, will over predict conversion by 25 to 30%.

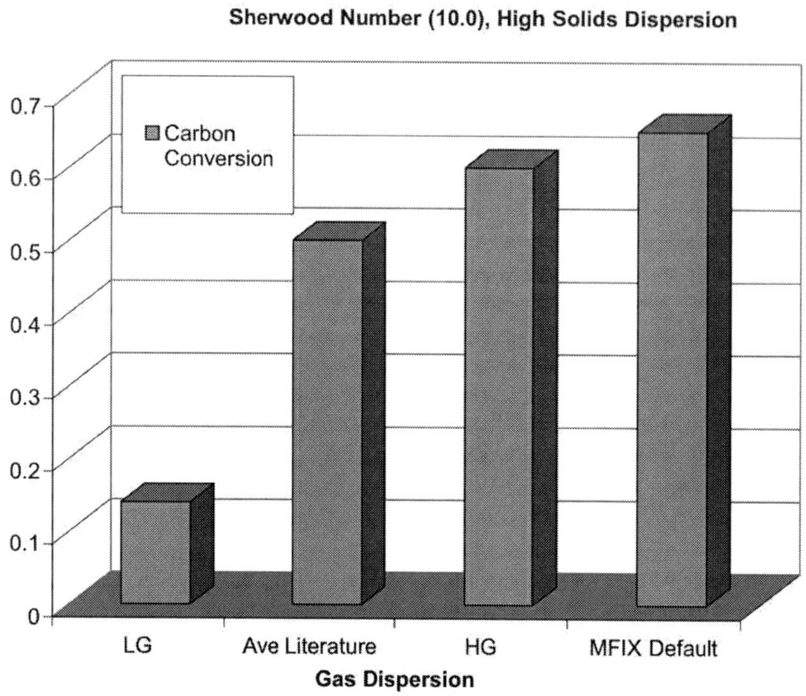

Figure 4 Carbon conversion for various gas dispersion coefficients.

To understand the sensitivity of the carbon conversion as a function of the Sherwood number simulations were conducted using high values for the solids dispersion coefficients and low values of the gas dispersion coefficients and then varying the Sherwood number. These results are given in Figure 5. For these kinetics, carbon oxidation to CO at a nominal temperature of 1144K, the Sherwood number has very little impact on the conversion for value greater than 0.5. Recent experimental data by Breault 2005b indicates that the Sherwood number is greater than 0.5 and is approximated closely by the Gunn correlation used by MFIX.

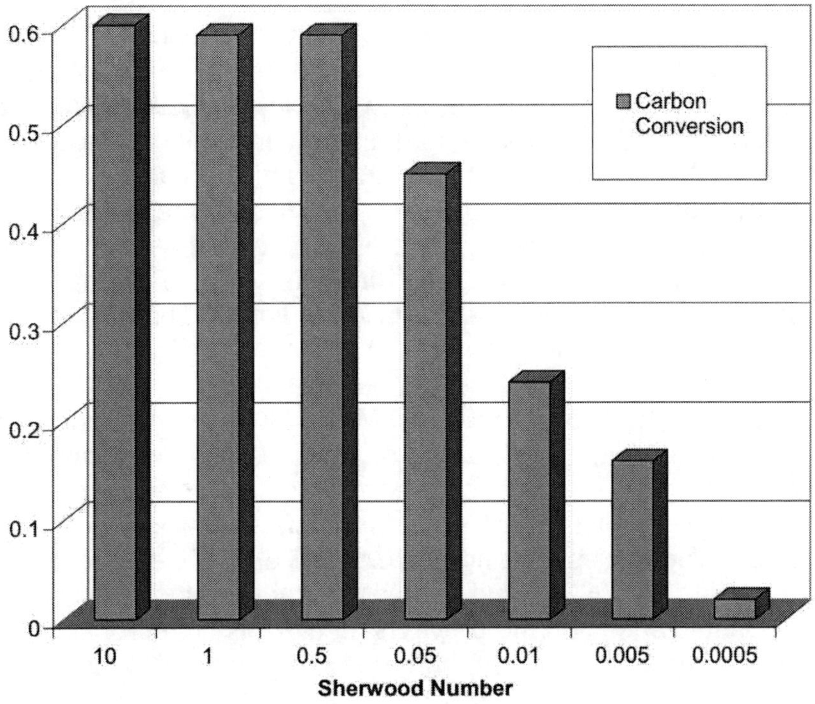

Figure 5 Carbon conversion for HS and LG as a function of the Sherwood number.

In order to determine whether the radial or axial dispersion coefficient was the dominant contribution to the carbon conversion two simulations were conducted. For these simulations the Sherwood number was determined using Gunn's expression (default in MFIX). The results given in Figures 2 and 3 showed that the gas dispersion coefficients had the dominant impact on carbon conversion and because of this the solids axial and radial dispersion coefficients were fixed at 5.0 and 0.5 respectively. The first simulation used a high axial gas dispersion coefficient of 5.0 and a low radial gas dispersion coefficient of 0.0005. The other simulation considered a low axial gas dispersion coefficient of 0.05 and a high radial gas dispersion coefficient of 0.05. The initial conditions and boundary conditions used in these simulations were identical to previous simulations reported earlier. The results showed for the high axial and low radial gas dispersion coefficients the carbon conversion was 15% and for the low axial and high radial gas dispersion coefficient simulation the carbon conversion was 52%. When comparing these values to earlier results it is clear that the sensitivity of the carbon conversion is mainly attributed to the choice of the axial gas dispersion coefficient.

Conclusion

Clearly, the results presented in figures 2-4 show the models predictions are sensitive to both the value of the Sherwood number and the value of the gas/solids dispersion coefficient. The results show that low gas dispersion and high solids dispersion produce the greatest carbon conversion independent of the Sherwood number. The results indicate that the gas dispersion has a dominant effect on the conversion while low or high values for the solids dispersion has secondary effects however, their effects do increase for the high Sherwood and high gas dispersion simulation. The results also show that the gas axial dispersion coefficient plays a dominant role in the carbon conversion compared to the gas radial dispersion coefficient.

In order to evaluate the default models in MFIX (Gunns expression for the Sherwood number, zero solids dispersion, and a gas dispersion coefficient based in part on a turbulent Schmidt number) a simulation was conducted using these default values. The values predicted for the mass transfer coefficients using Gunn's expression showed values ranging between 3.8 and 4.5, an average axial gas dispersion value of 0.00007 and an average radial dispersion value of 0.0025. With these values based on the previous results it is not surprising that MFIX predicted slightly higher conversion than the HS_LG simulation in Figure 3.

Nomenclature

CCPI	Clean Coal Power Initiative
CFB	Circulating Fluidized Bed F_{gs}
d_p	particle diameter (microns)
D_{O_2}	oxygen diffusivity (cm^2/s)
D_e	ash layer diffusivity (cm^2/s)
$\vec{g}$	force due to gravity (cm/s^2)
NETL	National Energy Technology Laboratory
P_g	pressure in the gas phase (Pa)
PSDF	Power Systems Development Facility
R	rate of production of the nth gas/solids species (g/cm^3s
$\overline{\overline{S}}_s$	solids phase stress tensor (Pa)
Sh	Sherwood number
Sc	Schmidt number
T_g	gas temperature(K)
T_s	solids temperature (K)
$\vec{v}_m$	gas/solids velocity vector (cm/s)
X_{mn}	n^{th} gas/solids species mass fraction
X^0	initial value of X_{mn}

Greek Symbols

ε_m gas/solids void fraction

μ_g gas viscosity (g/cms)

ρ_m gas/solids density (g/cm^3)

$\overline{\overline{\tau}}_g$ gas phase stress tensor (Pa)

References

Breault, R., "A Review of Gas-Solid Dispersion and Mass Transfer Coefficient Correlations in Circulating Fluidized Beds", AIChE National Meeting, Austin TX, Nov. 2004.

Guenther, C., M. Syamlal, L. Shadle, and C. Ludlow, "A Numerical Investigation of an Industrial Scale Gas-Solids CFB", Circulating Fluidized Bed Technology VII, J. grace, J. Zhu, and H. de Lasa Eds., Canadian Society for Chemical Engineering, Ontario, Canada (2002), pp. 483-496.

Gidaspow, D., R. Bezburuah and J. Ding, "Hydrodynamics of Circulating Fluidized Beds: Kinetic Theory Approach", Fluidization VII, O.E. Potter and D. J. Nicklin, Eds., Engineering Foundation, New York (1992), pp. 75-82.

Tsuo, Y. and D. Gidaspow, "Computational Flow Patterns in Circulating Fluidized Beds", AIChE J., **36**, 885-896, (1990).

Guenther, C., 'Brien, and M. Syamlal, "A Numerical Model of Silane Pyrolysis in a Gas-Solids Fluidized Bed", International Conference on Multiphase Flow, New Orleans, Louisiana, May 27-June 1, 2001.

Samuelbsberg, A., "Modeling and Simulation of Fluidized Bed Reactors", Ph.D. Thesis, Norwegian Institute of Technology, University of Trondheim, 1994.

Guenther, C., M. Syamlal, J. Longanbach, and P. Smith, "CFD Modeling of a Transport Gasifier Part II", Twentieth Annual Pittsburgh Coal Conference, September 15-19, 2003a Pittsburgh, Pennsylvania.

Guenther, C., M. Syamlal, J. Longanbach, and P. Smith, "Two-Fluid Model of an Industrial Scale Gasifier", AIChE Annual Meeting, November 16-21, 2003b San Francisco, California.

Syamlal, M., W. Rogers and T. O'Brien, "MFIX Documentation: Theory Guide", Tech. Note, DOE/METC-94/1004(1993).

Syamlal, M., "MFIX Documentation Numerical Technique", EG\&G Tech. Report,

DOE/MC31346-5824(1998).

Guenther, C. and M. Syamlal, "The Effect of Numerical Diffusion on Isolated Bubbles in a Gas-Solid Fluidized Bed", Powder Technology, **116**, 142-154 (2001).

Gunn, D., "Transfer of Heat or Mass to Particles in Fixed and Fluidized Beds", Int. J. Heat Mass Transfer, **21**, 467-476 (1978).

Desai, P. and C. Wen, "Computer Modeling of the MERC Fixed Bed Gasifier", MERC/CR-78/3, 1978.

Breault, R., C. Ludlow, and P. Yue, "Cluster Particle Number and Granular Temperature for Cork Particles at the Wall in the Riser of a CFB", Powder Technology, **149**, 68-77, (2005a).

Breault, R., "Mass Transfer in Circulating Fluidized Beds", AIChE Spring Meeting, Atlanta, GA, April, 2005b

Mass Transfer in Circulating Fluidized Beds

Ronald W. Breault[1] and Rex Compston
National Energy Technology Center
U.S. Department of Energy
3610 Collins Ferry Road, Morgantown, WV 26507-0880

Abstract

It is increasingly more important to be able to predict the conversion and yield of reactant species given the ever rising cost of the reactants and the ever decreasing acceptable level of effluent contaminants. As such, the development and use of predictive models for the reactors is necessary for most processes today. These models all take into account in some manner the interphase mass transfer. The model developer, unless equipped with specific experimentally based empirical correlations for the reactor system under consideration, is required to go to the open literature to obtain correlations for the mass transfer coefficient between the solid and gas phases. This is a difficult task at present since these literature values differ by up to 7 orders of magnitude. This paper presents a summary of the work in these areas as found in the literature dating back to 1949. The wide spread in the existing literature data is credited to flow regime differences in the reported literature and no distinction was made between test data taken in the acceleration zone and the fully developed zone.

Experiments are being conducted to carefully measure the dispersion and mass transfer in the fully developed region and in the acceleration region. In addition, these data are being taken in the turbulent, fast fluid bed and in the core-annular regimes. Data reflecting the findings will be presented and compared to the literature values.

Background

A major goal of the system designers and modelers is to be able to simulate the complex behavior in gas-solid chemically reacting flows in circulating fluidized bed systems. To support the mathematical simulation codes, research is being conducted on the hydrodynamics in these advanced gas-solids contactors, namely, circulating fluidized beds and transport reactors under atmospheric and pressurized conditions using various solids. The experimental data on the hydrodynamics of these flows provides crucial information to the modelers, providing them with experimental data on the hydrodynamic behavior of the gas and solids to which they can tune the models. To date, only one part of the simulation is rigorously being tuned with the experimental data, that being the overall flow field. There has been some effort to predict conversion in combustors and gasifiers. In these efforts, the performance has been modeled to predict the conversion using the pre-exponential factor in the rate expression as the adjustable tuning parameter. These are essentially one point CURVE fits and any result that passes through the experimental conversion would provide an acceptable result. This is overly simplistic; the simulation of combustors, gasifiers and the likes requires information on both gas and solids species movements and the interphase transport.

The development of Eulerian-Eulerian simulation codes such as MFIX to predict the performance of circulating fluidized beds and transport reactors to applications ranging from

[1] Author to whom communications should be addressed.

combustion and gasification to pollution control and chemicals production requires the existence of good theoretically based models to predict the interphase mass transport between reacting components and the dispersion of the reactants throughout the flow field within the reactors. MFIX uses the following equations for the gas and solids species continuity

$$\frac{\partial}{\partial t}(\varepsilon_g \rho_g x_{g,n}) + \nabla \bullet (\varepsilon_g \rho_g x_{g,n} u_g) = \nabla \bullet \mathcal{D}_{g,n} \nabla x_{g,n} + R_{g,n} \quad (1)$$

$$\frac{\partial}{\partial t}(\varepsilon_{s,m} \rho_{s,m} x_{s,m,n}) + \nabla \bullet (\varepsilon_{s,m} \rho_{s,m} x_{s,m,n} u_{s,m}) = \nabla \bullet \mathcal{D}_{s,m,n} \nabla x_{s,m,n} + R_{s,m,n} \quad (2)$$

To use the above equations, knowledge of the dispersion coefficients $D_{g,n}$ and $D_{s,m,n}$ is needed as well as the relationship for the interphase mass transfer coefficient, k_d, which is buried in the reaction terms, $R_{g,n}$ and $R_{s,m,n}$. The terminology in Equations 1 and 2 is taken directly from the MFIX document (Syamlal). Simulation codes such as MFIX allow for the dispersion coefficients to differ both directionally, spatially and by component.

The interphase mass transfer coefficient appears in the relationship for the reaction, R and is given in Equation 3 for a shrinking particle first order reaction with respect to the gas between gas species, j, and solid species, i.

$$R = \frac{-6(1-\varepsilon)C_j}{d_{p,i}\left(\frac{1}{k_c} + \frac{1}{k_d}\right)} \quad (3)$$

where k_c is the chemical reaction rate constant and k_d is the interphase mass transfer coefficient. In order to predict the performance conversion for a particular reaction in a CFB, values for the dispersion coefficients and the interphase mass transfer coefficient need to be obtained from the literature.

Literature Review

Experimental Facilities Used in Dispersion Investigations

A description of the operating conditions and experimental facilities for the review of the mass transfer data is presented in Table 4.

Table 1 Conditions and Dimensions for Mass Transfer Experimental Investigations

Author	d (m)	h (m)	T (K)	P (bar)	dp (um)	ρs (kg/m3)	Gs (kg/m2s)	Ug (m/s)
Li	0.072	3	ambient	ambient	300	1200	6 & 12	1.86 & 2.12
Subbarao	0.025	1.05	ambient	ambient	196 & 390	2500	~15 to 60	4, 5 & 6
Bolland	0.411	8.5	333	ambient	134	2500	31 to 53	5.6 to 7.2
Kalil	0.09525	0.3	ambient	ambient	685 to 2000	800 to 10000	N/A	N/A
Resnick	.022 to .044	0	ambient	ambient	3 to 35 mesh	0	N/A	N/A
Venderbosch	0.015	1.17	750	ambient	65	1375	5 to 40	2.5 to 4.5

Mass Transfer Coefficient Correlations

Six investigators have reported correlations for the interphase mass transfer coefficient in fluidized systems from research spanning 50 years. These are presented in Figure 4. There is little agreement between the reported results with values of the Sherwood number ranging from a low of 10^{-5} reported by Bolland to a high of 200 as reported by Kalil.

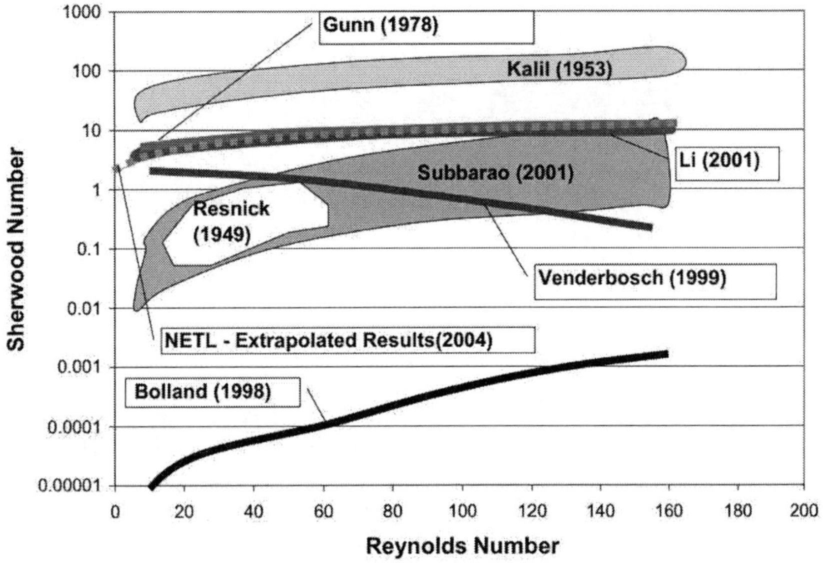

Figure 4 Mass Transfer Coefficients

Looking at the specific research that was conducted, it is possible to reduce this span by a couple orders of magnitude. Bolland conducted his research utilizing an ozone decomposition reaction. Ozone, upon first thought seems to be a good tracer to use, after all it decomposes to oxygen. Ozone decomposition is also highly dependent upon the moisture concentration in that gas and as no moisture data was taken and no effort made to control the moisture content during Bolland's test, his reported correlation for the Sherwood number is highly suspect. Eliminating this correlation from the analysis tightens the data range substantially, now being 0.01 to 200. The work of Kalil was conducted in a bubbling fluidized bed. He reports that the bed voidage is an important factor in the correlation of the Sherwood number. However, the application of this correlation to the riser of a CFB where voidages are on the order of 0.8 and higher likely introduces significant error in the estimate of the Sherwood number as the voidage in Kalil's work was on the order of 0.6. Thus, it can be expected for the Sherwood number to fall between 0.01 and 10. This is still a significant difference, but no other factors could be identified which suggest that any of the other correlations should be discarded.

Table 2 Mass Transfer Coefficient Correlations

Author (PI)	**Model/Data Range**
Li	$Sh = 2\varepsilon + 0.69(\frac{\mathrm{Re}_{sl}}{\varepsilon})^{0.5} Sc^{0.3}$
Subbarao	$Sh = 8.314 * 10^{-5}\left(\frac{U_o \rho_p}{G_s}\right)^{1.43}$
Bolland	$10^{-5} < Sh < 10^{-3}$
Zenenhoven	$Sh = 2\varepsilon + 0.89\,\mathrm{Re}_p^{0.5}\, Sc^{0.33}$
Kalil	$Sh = 1.77\left(\frac{\mathrm{Re}\,\varepsilon}{1-\varepsilon}\right)^{0.56} Sc^{0.33}$
Resnick	Graphical J factor
Venderbosch	Sh=Sh(Uo, solids fraction) Sh decreases with solids fraction and increase with Uo
Gunn	$Sh=(7-10\varepsilon+5\varepsilon^2)(1+0.7\mathrm{Re}_p^{02} Sc^{033})+(1.33-24\varepsilon+1.2\varepsilon^2)\mathrm{Re}_p^{07} Sc^{033}$

Discussion of Literature

The first two questions that arise with respect to understanding the wide level of discrepancies in the reported literature values for the mass transfer coefficients are:

- Were all the researchers working in the same flow regime as the solids and gas behavior is significantly different in the various regimes?
- Were all the test sections in a region of the riser well away from the effects of the entrance and exit, that is; was the flow regime fully developed?

The answers to these questions coupled to answering similar questions by the design/application engineer will help to eliminate some of the correlations presented in the literature and provide smaller ranges for these important transport properties.

Flow regime definitions have been developed to both qualify and quantify the solids and gas behavior in the risers of circulating fluidized beds and transport reactors. Due to the complexity of the system, different definitions have been developed by different investigators. Papers from Grace, Kato, Mei, Rhodes and Shadle as well as discussions with Monazam have been analyzed to develop a qualitative picture of the potential flow regimes that the mass transfer and dispersion researchers presented above were likely to encounter. This qualitative picture is presented in Figure 5.

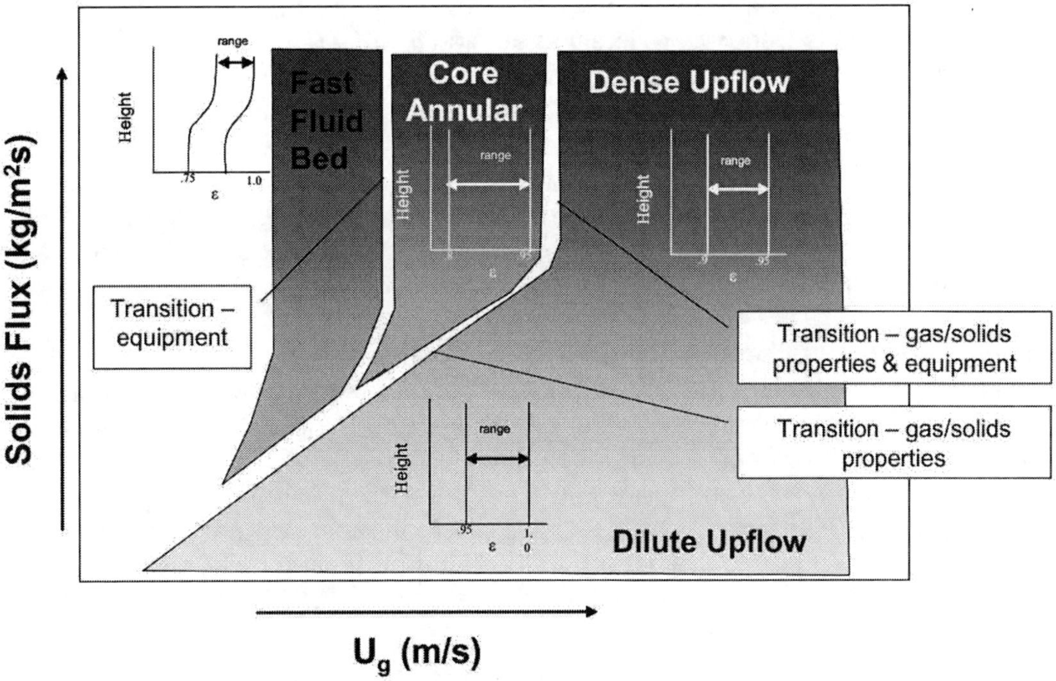

Figure 5 Flow Regimes in Circulating Fluidized Systems

The definition of a flow regime takes, at a minimum, 3 variables; solids flux, gas velocity and voidage (pressure) profile. Four regimes are presented in the figure: Fast fluid bed, Core annular, Dilute upflow and Dense upflow, although there is controversy as to whether or not the fast fluid bed regime and the core annual regime are the same as one definition simply states the co-existence of a dilute region and a dense region with the dilute region being above the dense region (denoted fast fluid bed in figure) or the dilute region being surrounded by a dense region (denoted core annular in figure). The transition from dilute up flow to either the fast fluid bed regime or the core annular regime depends upon the gas and solids properties. It is in the regions of this transition that most researchers work based upon physical limitations of their equipment. Therefore, depending upon what voidage profile existed in each of the experimental investigations, the flow regime may differ significantly.

Dilute upflow, is the simplest of the flow regimes. It consists of solids uniformly dispersed across the riser moving upward. At a particular gas velocity, increasing the solids flux decreases the voidage so as for the voidage to be reduced from 1 to about 0.95. Now depending upon the particle type and equipment a transition may be reached. At high gas velocities, no transition is reached and there is a gradual transformation from dilute upflow to dense upflow. At lower velocities, however, a transition is reached in which the riser solids profile significantly changes into one of two distinct patterns, fast fluid bed or core annular. Which of these two regimes that is reached, depends not only on the gas velocity but also on the equipment.

The fast fluid bed regime, as stated above, has a dense bottom zone and a dilute upper zone. By changing the solids flux, the solids in the upper region have collapsed into a dense lower region. In the dense region, solids flow downward as well as upward. Now clearly, the dispersion and the mass transfer are likely to be significantly different in these two regimes and significantly different with height. So depending on where the experimental measurements were taken is also likely to significantly affect the results. A similar phenomenon is obtained at slightly larger gas velocities with a transition to core annular flow, providing the equipment supports a core annular flow regime. In this circumstance, the uniformly distributed solids collapse to the wall and provide significant downflow. Again, this is significantly different behavior than dilute upflow and is likely to have significantly different values for dispersion coefficients and mass transfer coefficients.

As noted above, whether or not the tests were conducted in a fully developed flow field is also likely to affect the results of the work and contribute to the wide spread in the reported literature values for the mass transfer coefficient and the dispersion coefficients. Recently, Monazam has developed a correlation to predict the length of the acceleration zone. This work was conducted in a 0.3 m facility with a height of 16 m using 65□m glass beads, 806 □m cork particles and 230□m coke particles. Using the Monazam correlation to predict the acceleration length and comparing that length to the reported testing height reveals that about 50% of the data reported in the literature surveyed was obtained within the acceleration zone.

Experimental Methods

The test unit configuration is shown in Figure 1 and described by Shadle et al (2002). The solids enter the riser from a side port 0.23-m in diameter and 0.27-m above the gas distributor. Solids exit the riser through a 0.20-m port at 90° about 1.2-m below the top of the riser at a point 15.45-m above the solids entry location (centerline to centerline). Riser velocities were corrected for temperature and pressure as measured at the base of the riser. The air's relative humidity was maintained at 20% to minimize effects of static charge building up on the solids. The details of the facility, can be found elsewhere (Ludlow et al. (2002) and Shadle et al.(2002)).

Experiments were carried out by inserting a naphthalene sphere (moth ball) into the riser for a fixed time increment by holding the mothball in the probe shown in Figure 5. With the mothball securely clamped in the probe, it could be inserted and removed at will though a series of ball valves. The weight loss divided by the exposed time provided the mass transfer rate data needed to calculate the Sherwood number.

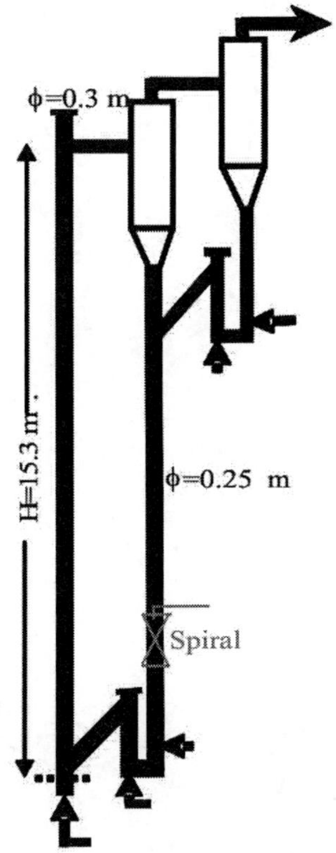

Figure 4 FBC Line Sketch

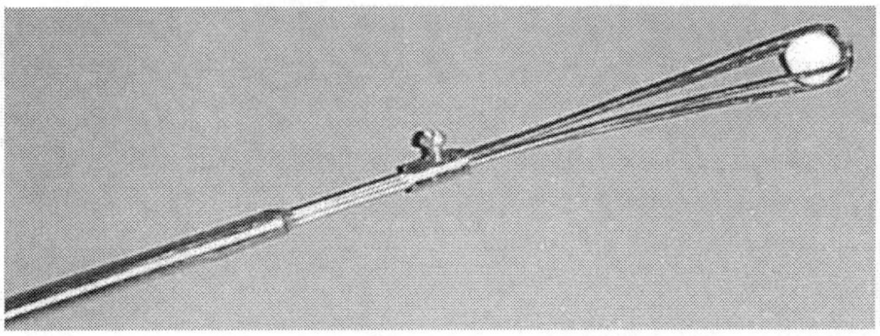

Figure 5 Mass Transfer Probe

Experiments Results

Experiments were first conducted in air, that is without the complicating effect of circulating solids, to confirm the technique for determining the Sherwood number. Figure 6 is a plot of the data as Sherwood number versus Reynolds number to ½ power. While, in general, the agreement is reasonable with the prediction based on the literature, the four data points with non-zero Reynolds numbers have Sherwood numbers that fall below the classic correlation for mass transfer from a single sphere in a fluid as given by

$$Sh = 2 + 0.6\,\mathrm{Re}^{0.5}\,Sc^{0.3}$$ **Equation 3**

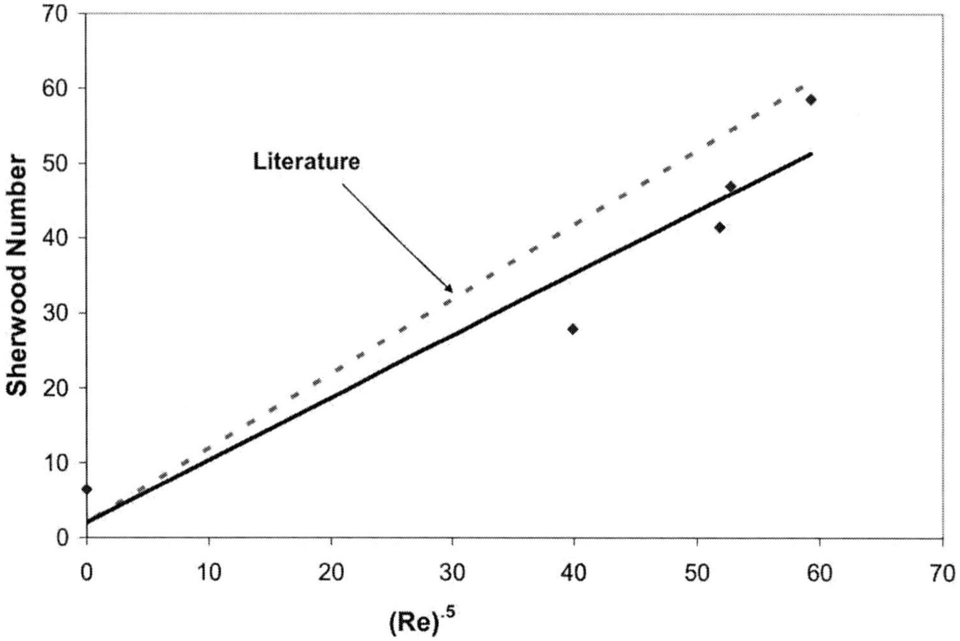

Figure 6. Single Particle Sherwood Number Vs. Reynolds Number to the ½ power

The likely cause of this deviation is the calculation of the mass transfer area. This area is calculated from the particle weight and density, assuming sphericity, and subtracting the area blocked by the probe fingers that hold the particle. Underestimating this blocking area, overestimates the mass transfer area and subsequently under predicts the Sherwood number. However, these values are within 25%, a reasonable deviation for mass transfer data. The value of the Sherwood number for the data with Reynolds number equal to zero is on the high side. This is likely to the assumption that the velocity was zero, that fact is, that the room was not free from convective currents and this movement increased the measured Sherwood number.

Expecting the mass transfer to increase and subsequently the Sherwood number when solids were introduced to the into the system led to great surprise when mass transfer as represented by the Sherwood number decreased as shown in Figure 7. The introduction of solids reduced the Sherwood number by a factor of about 2, from about 45 to about 22. It was expected that the solids would scrub the surface/boundary layer and increase the Sherwood number. However, that simple view did not account for the significant changes to the gas velocity profile, that is the local gas velocity, due to the introduction of the solids. The solids, even at the very dilute condition of 12,000 lb/hr of solids and a riser superficial gas velocity of 16.8 ft/s had solids flowing down at the wall. The velocity and solids fraction were measured with a fiber optic light probe to be about -0.1 ft/s and 1% solids for the case with 12,000 lb/hr of solids and about -1.0 ft/s and 10% solids for the case with 80,000 lb/hr. Substitution of velocities of this magnitude into Equation 3 gives values near the magnitude of the experimental data.

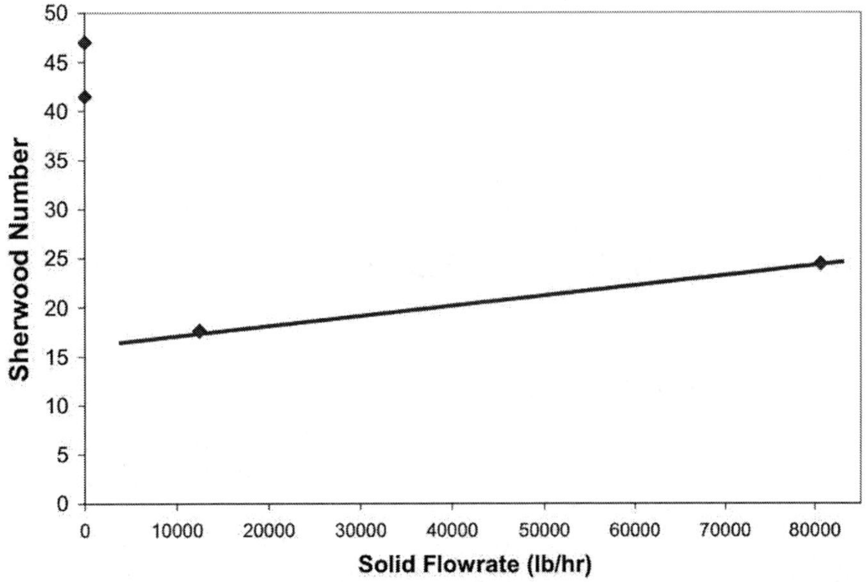

Figure 7. Effect of Solids Flow on Sherwood Numbers

The same data are plotted in Figure 8 against the Reynolds number to the ½ power. The data are then compared to the Gunn correlation as presented in Table 2. The gross assessment of the correlation to represent the data is that the correlation is reasonable, providing Sherwood numbers that are within acceptable error limits for mass transfer. The correlation appears to do better for higher Reynolds numbers than for low. In the CFB and transport reactor, the Reynolds number is more likely to be less than 100. In this region, while the Gunn correlation provides a better estimate of the Sherwood number additional data in this region could provide some useful information.

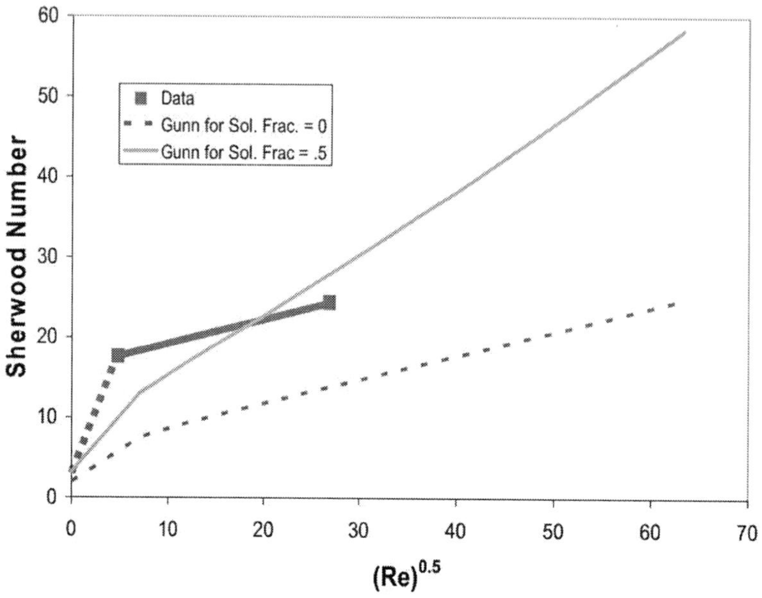

Figure 8. Comparison of Data with Gunn Correlation

Conclusion

Clearly the correlations presented by the researchers herein are not universally applicable. At present, there is no basis for recommending any of these correlations in favor of the others. Based upon the limited data, either the Gunn correlation or the Li correlation provide reasonable estimates for the Sherwood number in a CFB or transport reactor. The reasons for the wide discrepancy in the reported literature are likely due to the fluidization flow regime in which the researcher was working. Research needs to be conducted in a systematic manner across the operating regime map to determine these crucial transport properties. Studies conducted within the acceleration zone and in the fully developed zone needed to be conducted and reported for use within the respective zones. In some hardware configurations, the acceleration zone may be in excess of 50 percent of the bed riser height.

Nomenclature

C	Concentration, mol/m^3
d	diameter, m
G	Gas flux, kg/m^2s
Gs	Solids flux, kg/m2s
k	rate constant or coefficient, m/s
N	particle count
R	Reaction Rate, mol/s
Re	Reynolds Number
Sc	Schmidt Number
Sh	Sherwood Number
t	time, s

U Velocity, m/s
x mole fraction

Greek
ε Voidage
μ Viscosity, kg/ms^2
ρ Density, kg/m^3

Subscripts
0 initial
c chemical
d mass transfer
g gas
i index
m index
n index
p particle
s solid
sl slip

References

Bolland, O., "Describing Mass transfer in Circulating Fluidized Beds by Ozone Decomposition", REPORT 1998:02

Geldart, D. "Types of gas fluidization", 1973 *Powder Technology* 7: 285-292.

Grace, J. R., K. S. Lim, and J. X. Zhu, "Hydrodynamics of gas-Solid Fluidization", Itt. J. Multiphase Flow, Vol 12, Suppl. Pp 141-193, 1995

Gunn, D. J.,"Transfer of Heat or Mass to Particles in Fixed and Fluidized Beds", Int. J. of Heat and Mass Transfer, 21, 467-476 (1978)

Kato, K., G. Xu, K. Nomura and N. Nakagawa, "Hydrodynamic dependence on the riser diameter for different particles in circulating fluidized beds", Powder Technology 113 (2000) 80-87

Kalil, J., J. C. Chu and W. A. Wetteroth, "Mass Transfer in a Fluidized Bed", CEP 49. No. 3, pg 141, 1953

Li, J. and L. Wang, "Concentration Distributions During Mass Transfer in Circulating Fluidized Beds", 7th ICCFB 2002, Niagra Falls

Mei, J. S., E. R. Monazam and L. J. Shadle, "The Mapping of Flow Regime for a Light Material: Cork", 17th ICFBC, May 18-21, 2003, Jacksonville, FL.

Monazam, E. R., personal communication regarding acceleration and fully developed zones.

Resnick, W. and R. R. White, "Mass Transfer in Systems of Gas and Fluidized Solids", CEP 45, No 3., pg 377, 1949

Rhodes M. J., X. S. Wang, "Defining Fast Fluidization", Fluidization IX, 1998, pp133

Shadle, L. J., J. C. Ludlow, J. S. Mei, and C. Guenther, "Circulating Fluidized Bed Technology for Advanced Power Systems", 7th ICCFB 2002, Niagra Falls

Subbarao, D. and S. Gambhir, "Gas Particle Mass Transfer in Risers", 7th ICCFB 2002, Niagra Falls

Syamlal, M., "MFIX Document Numerical Technique", Department of Energy Report DOE/MC31346-5824, Jan. 1998.

Venderbosch Mass Transfer and Influence of the local catalyst activity on the conversion in a Riser reactor Can J of Chem Eng V77, April 1999, pg 262

Zenenhoven, R. and M. Jarvinen, CFB reactors, Particle Slip Velocity and Particle/Turbulence Interactions", 7th ICCFB 2002, Niagra Falls

Zenenhoven, R. and M. Jarvinen, "CFB reactors, CFD and Particle/Turbulance Interactions, 4th ICMF 2001, New Orleans

Comparison of experiments and multiphase particle-in-cell simulations in a two-dimensional fluidized bed

Paul R. Tortora, Timothy J. O'Hern; Sandia National Laboratories
Ken Williams, Dale Snider; Arena-flow LLC

Abstract

Two-dimensional (2D) fluidized beds are sometimes used to allow examination of flows representative of visually opaque three-dimensional fluidized beds. This paper presents data acquired in a 2D fluidized bed consisting of air rising through a bed of glass beads. Data include pressure measurements along the height of the riser and particle image velocimetry (PIV) velocity measurements of bead movement at the wall. Turbulent statistics are extracted from the PIV data. A computational simulation using a multiphase particle-in-cell model is presented. The computational and experimental results are compared.

Introduction

The overall goal of this work is the formulation of computational models to predict and improve the behavior of industrial-scale fluidized bed processes. Specific applications being studied here include the Flexicoking™ and Fluid Coking™ processes (ExxonMobil Corp.) and titanium dioxide production (Millennium Chemicals Co.). To model these production reactors, new physics and phenomenological models are being implemented in the commercial computational fluid dynamics software package *Arena-flow®*. A pilot-scale fluidized bed is being constructed at Sandia National Laboratories to obtain experimental data for validation.

In this work, a small-scale 2D fluidized bed has been constructed to examine some fundamental flow phenomena that may be present in the larger experiment. 2D beds are often used in fundamental studies of fluidized beds because flow phenomena such as bubbles may be visually observed. The bed used here is 50 cm wide, 100 cm high (filled to 60 cm with particles) and 1.35 cm in depth. The particles used here are glass spheres ranging in size from 120 to 180 microns. Movies are recorded with digital cameras and PIV is performed on the images. *Arena-flow®* simulations of this geometry are also performed, making it a useful validation case as well. Experiments and simulations with a stream of air introduced from the side of the bed (a "side jet") were also performed.

Arena-flow®

Arena-flow® is computational fluid dynamics software developed specifically for dense-phase particle flows pertinent to diverse industrial applications. The numerical method treats the particle phase as both a continuum and as discrete particles. Particle properties are mapped to and from an Eulerian grid. While on the grid, continuum derivatives that treat the particle phase as a fluid are evaluated and then mapped back to the individual particles. The result of this procedure is a computational technique for multiphase flow that can accurately handle particle loadings ranging from dilute to dense with a distribution of particle materials and sizes. A detailed description of the numerical method is given by Snider (2001), and a brief description is given here.

The fluid (gas) phase is incompressible, and fluid and particle phases are isothermal at present, but under the current project gas-phase compressibility and energy equations are being implemented. The fluid or Eulerian variables are denoted by subscript *f*, and the particle or Lagrangian variables are denoted by subscript *p*. The continuity equation for the fluid with no interphase mass transfer is

$$\frac{\partial \theta_f}{\partial t} + \nabla \cdot (\theta_f \boldsymbol{u}_f) = 0 \tag{1}$$

where $\boldsymbol{u}_f$ is the fluid velocity and θ_f is the fluid volume fraction.

The momentum equation for the fluid is

$$\frac{\partial(\theta_f \boldsymbol{u}_f)}{\partial t} + \nabla \cdot (\theta_f \boldsymbol{u}_f \boldsymbol{u}_f) = \frac{-1}{\rho_f} \nabla p - \frac{1}{\rho_f} \boldsymbol{F} + \theta_f \boldsymbol{g} + \theta_f \mu_t \nabla^2 \boldsymbol{u}_f \tag{2}$$

where ρ_f is fluid density, *p* is fluid pressure and $\boldsymbol{g}$ is the gravitational acceleration. $\boldsymbol{F}$ is the rate of momentum exchange per volume between the fluid and particle phases. Following averaging methods in single phase flow, Eq. (2) is the Reynolds average momentum equation, where μ_t is the sum of the molecular viscosity and the turbulent eddy "viscosity". The momentum equation neglects $\mu \nabla(\nabla \cdot \boldsymbol{u})$ viscous terms which arise in two-phase flow (zero for single phase incompressible flow). A single equation length scale turbulence model from the Los Alamos Telluride program is used (Kothe et al., 2001).

The dynamics of the particle phase is described by a particle probability distribution function (Snider, 2001; Travis et al., 1976; Andrews and O'Rourke, 1996; Snider et al., 1998). The mass of each particle is constant in time (no mass transfer between particles or to the fluid), but particles may have a range of sizes and densities. A computational particle method is used to solve for the particle distribution. Particles are grouped into computational parcels termed "clouds", each containing n_p particles with identical density ρ_p, volume V_p and velocity $\boldsymbol{u}_p$ located at position $\boldsymbol{x}_p$. The particle acceleration is given by

$$\frac{d\boldsymbol{u}_p}{dt} = D_p(\boldsymbol{u}_f - \boldsymbol{u}_p) - \frac{1}{\rho_p} \nabla p + \boldsymbol{g} - \frac{1}{\theta_p \rho_p} \nabla \tau_p \tag{3}$$

The terms on the right are accelerations due to aerodynamic drag, fluid pressure gradient, gravity and gradient in the particle normal stress (collision).

The aerodynamic drag model is a modified Stokes equation:

$$D_p = C_d \frac{3}{8} \frac{\rho_f}{\rho_p} \frac{|\boldsymbol{u}_f - \boldsymbol{u}_p|}{\boldsymbol{r}} \tag{4}$$

where

$$C_d = \frac{24}{R_N}\left(\theta_f^{-2.65} + \frac{1}{6}R_N^{\frac{2}{3}}\theta_f^{-1.78}\right) \quad for \quad R_N < 1000$$

$$C_d = 0.44\theta_f^{-2.65} \quad for \quad R_N \geq 1000 \tag{5}$$

The Reynolds number is defined as

$$R_N = \frac{2\rho_f \left|\boldsymbol{u}_f - \boldsymbol{u}_p\right| r}{\mu_f} \tag{6}$$

where μ_f is the fluid molecular viscosity and r is the particle radius.

Collisions between particles are modeled using an isotropic interparticle stress where the off-diagonal elements of the stress tensor are neglected. The particle stress model used is modified from Harris and Crighton (1994) which has the same general form as the radial distribution function in the kinetic theory for dense particle flow (Jenkins and Savage, 1983):

$$\tau_p = \frac{P_s \theta_p^{\ 3}}{Max\left(\theta_{CP} - \theta_{p,0}\right) + \varepsilon} \tag{7}$$

where the constant P_s has units of pressure, θ_{CP} is the particle volume fraction at close packing, and the small number in the denominator ε, which is on the order of 10^{-7}, removes the singularity. The particle stress is unaffected by ε except when the volume fraction approaches or exceeds close-pack. The close-pack limit is somewhat arbitrary and depends on the size, shape and ordering of the particles. Allowing the particle volume fraction to slightly exceed close-pack is physically possible considering that shifting or rearranging of granular materials may occur. The ε allows a smoother numerical solution than bumping up against a singularity. The continuum particle stress equation depends only on the concentration of particles and neglects the size and velocity of particles. The continuum normal stress gradient is applied to the discrete particles as described by Snider (2001).

The continuum equations are approximated with finite volumes on a staggered grid. A first order upwind advection operator blended with a second order central difference advection operator is used, and a backward difference time operator is used. The implicit coupling of fluid and particles through the interphase momentum transfer is

$$F_{\xi}^{n+11} = \frac{1}{\Omega_{\xi}} \sum S_{p\xi} \left[D_p \left(\boldsymbol{u}_{fp}^{\ n+1} - \boldsymbol{u}_p^{\ n+1} \right) - \frac{1}{\rho_p} \nabla p_p \right] n_p m_p \tag{8}$$

where m_p is the mass of a particle, n_p is the number particles in a cloud, Ω_{ξ} is the cell volume, and ξ is the grid cell. The summation is over all particles and $S_{p\xi}$ is the interpolation operator from particle location to grid. The fluid velocity in Eq. (8), $\boldsymbol{u}_{fp}^{n+1}$, is at a particle position which requires an interpolation involving neighboring grid velocities. Similarly, the interpolated pressure gradient ∇p_p and particle stress gradient $\nabla \tau_p$ introduce neighbor quantities. The numerical approximation leads to a set of algebraic equations whose basic forms have been described extensively in publications and reviewed in books such as from Ferziger and Peric (1996), and are described by Snider (2001).

Conservation of volume is from the Lagrangian particles where particle locations are calculated from

$$x_p^{n+1} = x_p^n + \Delta t u_p^{n+1} \tag{9}$$

Because of the Lagrangian calculation, the conservation of volume (mass) is calculated without advection numerical diffusion. The particle volume mapped to the Eulerian grid is

$$\theta_{p\xi} = \frac{1}{\Omega_\xi} \frac{\sum^{n_p} m_p n_p S_{p\xi}}{\rho_p} \tag{10}$$

where Ω_ξ is the volume and ξ is the grid cell, and conservation of volume for all materials is enforced by ($\theta_p+\theta_f=1$). To insure continuity from the continuum conservation of volume, Eq. (1), a pressure/velocity error equation is formed from Eq. (1) and the momentum equations. A SIMPLE type solution is used to solve the pressure/velocity correction which satisfies continuity.

Arena-flow® solves transient, three-dimensional dense-phase problems and input geometry is from a CAD solid (3D) model. The grid follows the physical geometry, and because the particles are calculated in a Lagrangian frame of reference, all surfaces triangles and their normals are required to numerically keep particles within the grid. Particles impacting solid walls have an elastic restitution and can rebound from walls with user-specified spectral-to-diffuse bounce.

Previous experimental work

2D fluidized beds are often used to study flow phenomena in gas-solid flows, to give insight into flow dynamics in industrial-scale three-dimensional beds. Two-dimensional beds are typically rectangular and narrow (on the order of 1 cm), so that flow phenomena (such as bubbles) may be observed visually. Researchers have examined flow phenomena in two-dimensional fluidized beds such as pressure fluctuations, bubble size, bubble velocity, particle velocity, and particle segregation. Measurements have also been compared to theoretical relations and computational simulations.

Kuipers et al. (1991) filmed the growth of a bubble in a two-dimensional fluidized bed and compared the results with theoretical predictions and computational simulations. Air flowed into the bed through a porous distributor at its base, and bubbles were formed with air flowing through a central orifice. Similarly, Bokkers et al. (2004) filmed bubbles in a two-dimensional fluidized bed with a central orifice and compared the results to computation simulations. These investigators also performed PIV and segregation experiments with different-colored particles, comparing the results to computational simulations.

Daw et al. (2000) used pressure measurements in a two-dimensional fluidized bed to characterize bubble behavior. Cross-correlations of the pressure data were used to estimate bubble rise velocities, and slow oscillations (on the order of 100 seconds) were also observed in the pressure data. Kage et al. (2004) measured bubble diameters and velocities in a two-dimensional fluidized bed with digital cameras. Additionally, they examined a vibrating fluidized bed, and compared bubble diameters and velocities with theoretical relations.

Other investigators describe automated image analysis techniques for measurements in two-dimensional fluidized beds. Shen et al. (2004) describe automated techniques to determine bubble diameter, bubble velocity, bubble density, and gas throughflow, and compared measurements to theoretical predictions. Goldschmidt et al. (2003) used color photography and image analysis to distinguish different-colored particles from each other and from a colored background. They were able to describe the expansion dynamics (frequency and height of oscillations) for each particle species and measure segregation.

The behavior of side jets has also been studied, the penetration depth being of particular interest. Ariyapadi et al. (2003) used X-rays to image liquid and gas side jets and estimate their shape and penetration depth.

Experiments and general observations

The 2D bed used in this work is shown schematically in Figure 1. It is constructed of clear acrylic sections glued together. The plenum is attached to the top portion of the bed by a flange, so that it may be disassembled to insert different kinds of gas distributors. The distributor used in this work is an "air porous board" (Applied Plastics, Inc., 10 micron pore size). The particles used here are spherical glass beads ranging from 120 to 180 microns in diameter with a mean diameter of 150 microns. A micrograph is shown in Figure 2. The particle density is approximately 2660 kg/m^3 and they fall into Group B of the Geldart classification.

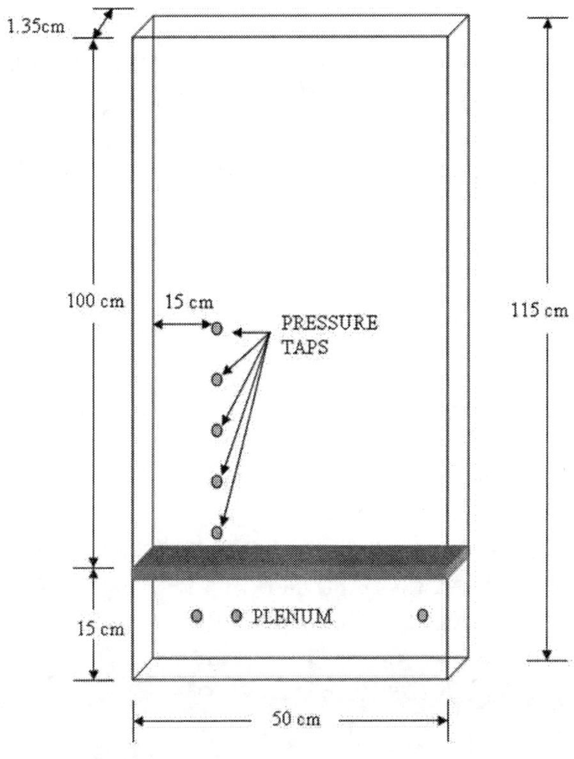

Figure 1: 2D bed schematic (inner dimensions)

Dry supply air is first fed into a venturi, which entrains humidified air (produced in a vessel of heated water through which air is bubbled). The air then flows past a humidity/temperature probe, through a needle valve used for throttling, and finally a flow meter which measures the flow rate in liters-per-minute (lpm). The relative humidity of the inlet air is typically 70%. The air then flows into the plenum at the base of the fluidized bed. All air lines are 1/4-inch tubing. The air is distributed into the bed through the distributor. The pressure drop across the distributor is approximately 0.017 psi/slpm.

Pressure taps are located in the plenum, 2 cm above the air porous board, and every 10 cm above that. Validyne pressure transducers are used to measure pressure and are monitored by Labview data acquisition software, which also records flow rate, temperature, and humidity of the inlet air, and atmospheric pressure and temperature. All quantities are measured at 100 Hz. Glass beads are prevented from contaminating the pressure transducers by sintered metal discs with a 10 micron pore size inserted into the fittings. Flow rates and pressures are monitored using Labview™ data acquisition software on a personal computer.

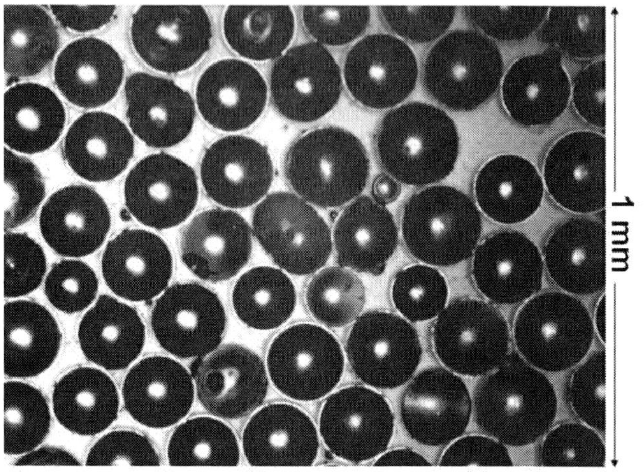

Figure 2: Glass beads

In a typical experiment the flow rate of air is increased by intervals. The throttling valve is opened as quickly as possible at each interval for as a rapid an increase in flow rate as possible. At each interval, the bed is allowed to operate for several minutes. Pressures are very stable up until the point when bubbling begins (at the minimum fluidization velocity u_{mf}). After bubbling begins, pressures are unstable and the flow can only be considered "quasi-steady": when pressure averaged over several seconds becomes constant, which usually occurs after approximately one minute of changing the air flow rate. Average pressures recorded at a number of flow rates are shown in Figure 3. This behavior is typical of fluidized beds, except that in this case the bed partially defluidized (at a flow rate of approximately 12 slpm).

Experiments were also performed with a side jet. The side jet was located 19 cm above the distributor in the narrow side of the bed. A single flow rate of 8.7 slpm was used in this work, flowing through a 0.76 cm diameter port, making the inlet velocity 3.2 m/s. Some general observations: With the bed defluidized (no air flowing through the distributor), starting the side jet first formed a pocket of air at the inlet which

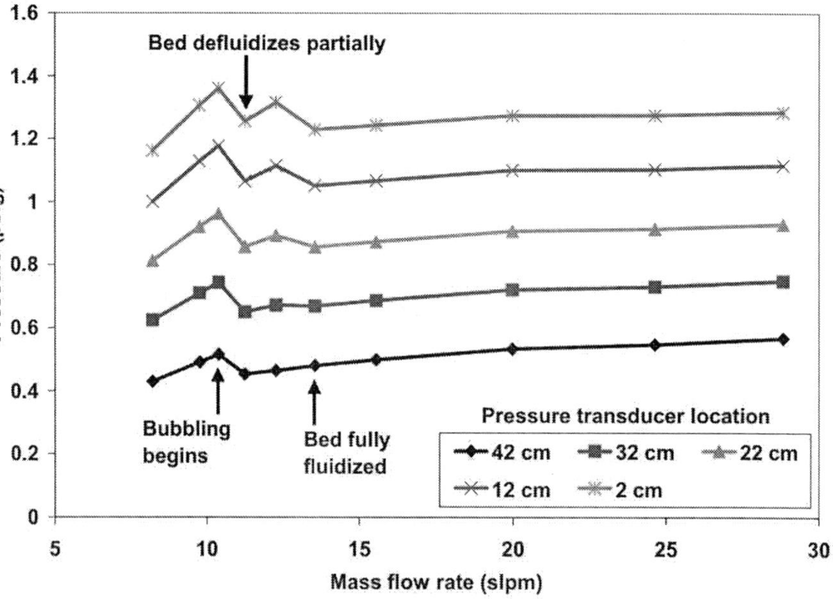

Figure 3: Fluidized bed pressures averaged over 10 seconds

remained stable. Tapping the bed allowed the pocket to grow and reach the surface of the bed after which air continuously bubbled from the jet to the bed surface. An image sequence of this flow is shown in Figure 4. The air traveled almost directly up from the port (very little penetration in the horizontal direction) and was not at all "jet-shaped". This remained the same when the jet was introduced into a bubbling bed. Other investigators have observed "jet-shaped" side jets (Ariyapadi et al., 2003) but they used much higher velocities. In the current experiments, the flow rate through the side jet was a modest fraction of the flow rate through the distributor and could not be increased with the current setup.

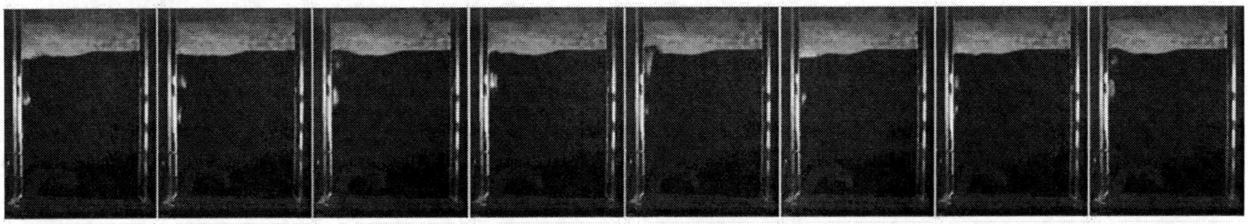

Figure 4: Image sequence of side jet (on the left) in quiescent bed (5 frames per second); bed dimensions are 50 cm wide by 60 cm high

Side-by-side comparison of experiments and simulations

An *Arena-flow*® simulation was performed with the following inputs: The inner dimensions of the bed are as mentioned above, with a 0.46 cm^2 circular hole at a height of 19 cm for the side jet. The particles range in size from 120 to 180 microns which was captured by the numerical formulation, their density is 2660 kg/m^3, and the close-pack volume fraction is 0.6. The bed is initially quiescent, then air begins to flow through the distributor with a velocity of 0.01235 m/s. This velocity corresponds to a mass flow rate of 19.4 slpm measured in the experiment converted to velocity, assuming that the density of the air at the distributor corresponds to the measured pressure through the ideal gas law. It is also assumes that the air flows through an area fraction of 0.2, i.e. the velocity stated above is 5 times the superficial gas velocity. The simulation is run for sufficient time to achieve a quasi-steady state, then the jet is introduced with a velocity of 3.18 m/s. This velocity corresponds to a mass flow rate of 7.6 slpm using a similar conversion as above with a area fraction of 1. The mesh is 1 cell thick in the narrow dimension of the bed (1.35 cm).

Figures 5 and 6 show image sequences of *Arena-flow*® simulations and experiments performed without and with the side jet operating. All were acquired when a quasi-steady state was reached in the simulations/experiments. The images from the experiments were illuminated from the back, so the only bubbles visible are those that spanned the depth of the bed (1.35 cm). The images from the simulations are a subset of particle clouds colored in grayscale corresponding to the solid volume fraction in each cell.

These images allow for some qualitative comparisons between the experiments and simulations. In the future, physical data will be extracted from both to make quantitative comparisons. It may be generally observed that the bubble sizes, velocities, and growth in both the experiments and simulations are very similar. The measure bed pressure compared nearly identically to the *Arena-flow*® computational results. The bubbles appear somewhat "rounder" in the simulations. Also, the side jet appears to penetrate further into the bed horizontally in the simulation than in the experiment. These differences may be due to the presentation of the data (light transmitted in the experiment vs. grayscale of solid volume fraction in the simulation) or a need to modify the physics in the simulation, or both. The true reasons for the difference between simulations and experiments will be determined by comparing further physical data from the two, and by performing parametric studies with *Arena-flow*®.

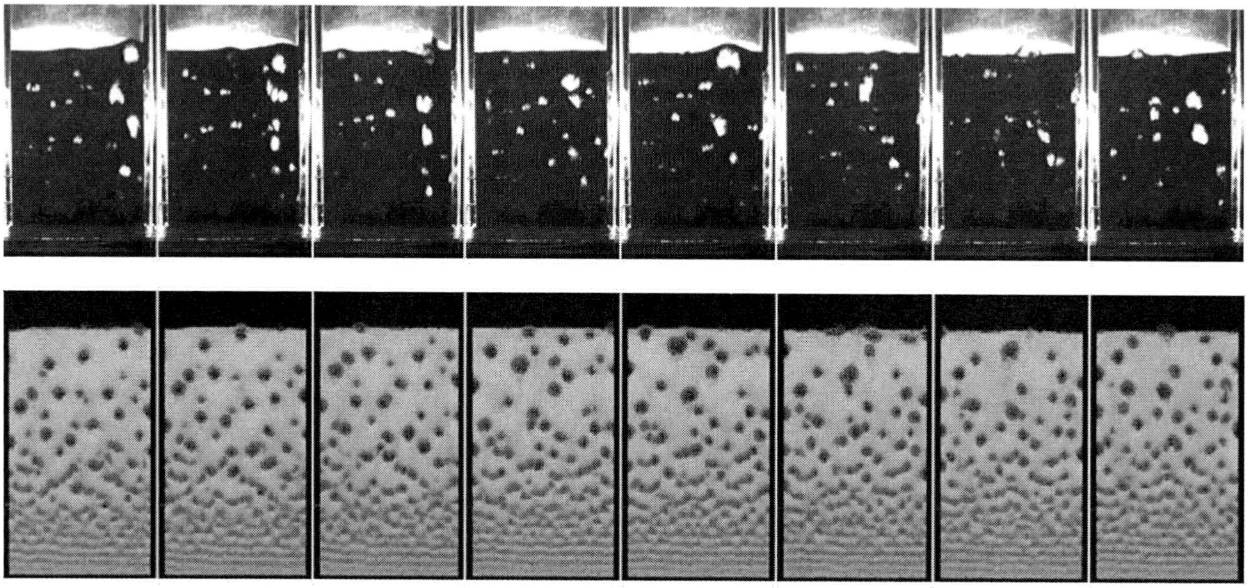

Figure 5: Image sequence of fluidized bed operating without side jet (5 frames per second)

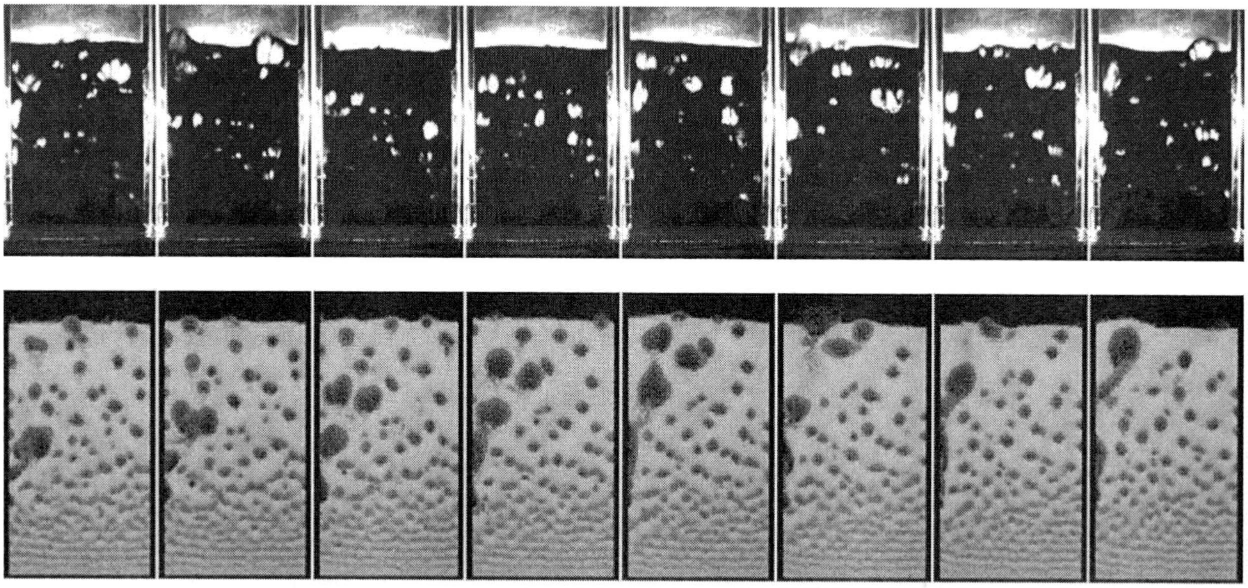

Figure 6: Image sequence of fluidized bed operating with side jet (5 frames per second); side jet is on the left

Particle-image velocimetry

PIV was performed by illuminating and filming the glass particles in a 5 cm square region. Separate tracer particles were not required because individual glass particles could be resolved and used as tracers. The bed was filmed at 500 frames per second. The software *PIV Sleuth* (Christensen et al., 2000) was used to interrogate and validate the images. An example of the vector field obtained (average vertical and horizontal velocity subtracted) is shown in Figure 7. The image was filmed in the approximate center of the bed (25 cm horizontal, 30 cm vertical). The region of outflow seen in the center of the image corresponds to a bubble. Statistics were performed on 20,000 images (average velocity and turbulence) but taking a running average shows that they do not converge. Therefore, more images are necessary to obtain the average flow field and turbulence statistics. Typically, this number of images is sufficient to obtain averages with PIV, but in this case the size and frequency of the bubbles represented large fluctuations compared to the time within which the images were acquired. These PIV images are therefore preliminary in nature.

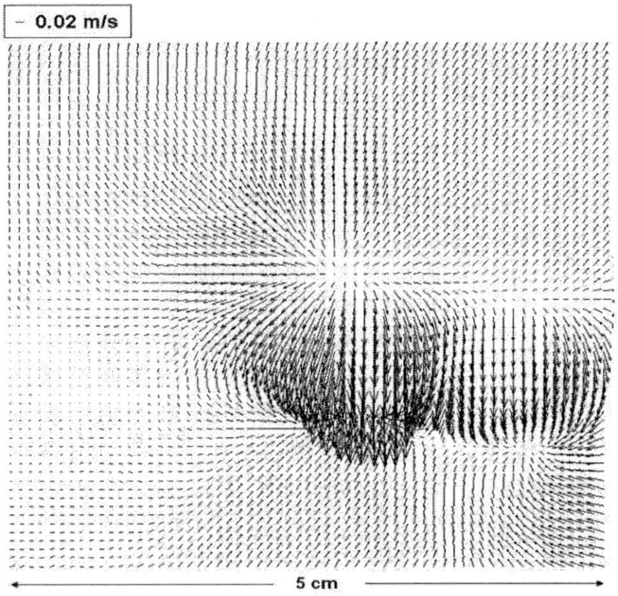

Figure 7: Example of PIV image (average vertical and horizontal velocity subtracted)

Conclusions

Side-by-side simulations and experiments in a 2D fluidized bed have been performed, with the goal of validating the software *Arena-flow®* for industrial-scale fluidized bed applications. The 2D bed is a small-scale experiment used to examine flow phenomena that will be present in a larger pilot-scale experiment, in which similar comparisons between experiments and simulations will also be performed. The image sequences shown here may be used to make qualitative comparisons, but in the future physical data will be extracted from both the experiments and simulations to make quantitative comparisons. This data will include pressure and flow rate data, and bubble characteristics (e.g., shape, size, and velocity). It has also been shown that PIV can be performed, but more images than are presented here are needed to obtain meaningful statistics.

Acknowledgements

The authors gratefully acknowledge the support of the Department of Energy Industrial Technologies Program. This work is funded under award number DE-FC36-04GO14153, Bill

Prymak, project officer. The skilled technical contributions of John Oelfke – technologist at Sandia National Laboratories – are also acknowledged. Kathryn Sy of Texas A&M University – summer intern at Sandia National Laboratories – developed and instrumented the first version of the 2D fluidized bed. Sandia is a multiprogram laboratory operated by Sandia Corporation, a Lockheed Martin Company, for the United States Department of Energy's National Nuclear Security Administration under contract DE-AC04-94AL85000.

References

Andrews, M.J., and O'Rourke, P.J., (1996), "The multiphase particle-in-cell (MP-PIC) method for dense particle flow," *International Journal of Multiphase Flow* **22**, 379-402.

Ariyapadi, S., Holdsworth, D.W., Norley, C.J.D., Berruti, F., and Briens, C., (2003), "Digital X-ray imaging technique to study the horizontal injection of gas-liquid jets into fluidized beds," *International Journal of Chemical Reactor Engineering* **1**, A56.

Bokkers, G.A., van Sint Annaland, M., and Kuipers, J.A.M., (2004), "Mixing and segregation in a bidisperse gas-solid fluidised bed: a numerical and experimental study," *Powder Technology* **140**, 176-186.

Christensen, K. T., Soloff, S. M., and Adrian, R. J., (2000), "PIV Sleuth - Integrated Particle Image Velocimetry Interrogation/Validation Software," TAM Report No. 943, University of Illinois.

Daw, C.S., Halow, J.S., Finney, C.E.A., and Nguyen, K., (2000), "Characterizing the hydrodynamics of bubbling fluidized beds with multivariate pressure measurements", *AIChE Annual Meeting*, Los Angeles, CA.

Ferziger, J.H., and Peric, M., Computational methods for fluid dynamics, Springer-Verlag, Berlin, Heidelberg (1996).

Goldschmidt, M.J.V., Link, J.M., Mellema, S., and Kuipers, J.A.M., (2003), "Digital image analysis measurements of bed expansion and segregation dynamics in dense gas-fluidised beds," *Powder Technology* **138**, 135-159.

Harris, S.E., and Crighton, D.G., (1994), "Solutions, solitary waves and voidage disturbances in gas-fluidized beds," *Journal of Fluid Mechanics* **266**, 243-276.

Jenkins, J.T., and Savage, S.B., (1983), "A theory for the rapid flow of identical smooth, nearly elastic spherical particles," *Journal of Fluid Mechanics* **130**, 67-93.

Kage, H., Funaoka, S., Kohno, H., Zhou, T., and Yamamura, M., (2004), "Effect of vibration addition on bubble size and velocity in two-dimensional fluidized bed" in Fluidization XI (ed. U. Arena, R. Chirone, M. Miccio, and P. Salatino), Engineering Conferences Foundation, New York.

Kothe, D.B. et al., (1995), “Computer simulation of metal casting process: A new approach,” Technical Report LALP-95-197, Los Alamos National Laboratory, Los Alamos, NM.

Kuipers, J.A.M., Prins, W., and van Swaaj, W.P.M., (1991), "Theoretical and experimental bubble formation at a single orifice in a two-dimensional gas-fluidized bed," *Chemical Engineering Science* **46**, 2881-2894.

Shen, L., Johnsson, F., and Leckner, B., (2004), "Digital image analysis of hydrodynamics two-dimensional bubbling fluidized beds," *Chemical Engineering Science* **59**, 2607-2617.

Snider, D.M., (2001), “An incompressible three dimensional multiphase particle-in-cell model for dense particle flows,” *Journal of Computational Physics* **170**, 523-549.

Snider, D.M., O’Rourke, P.J., and Andrews, M.J., (1998), “Sediment flow in inclined vessels calculated using multiphase particle-in-cell model for dense particle flow,” *International Journal of Multiphase Flow* **24**, 1359-1282.

Travis, J.R., Harlow, F.H., and Amsden, A.A., (1976), “Numerical calculation of two-phase flows”, *Nuclear Science and Engineering* **61**, 1-10.

120a Numerical Investigation of Water and Thermal Management in PEM Fuel Cells

Jinquan Xu, M. Yousuff Hussaini

A proton exchange membrane (PEM) fuel cell is an electrochemical device that converts chemical energy of hydrogen directly into electricity, by combining hydrogen and oxygen into water. Heat is released in the process. One of the technical challenges facing efficient PEM fuel cell development is water and thermal management. With a view to investigate optimal water and thermal management, an integrated non-isothermal multi-species, two-phase, porous medium, computational model for the entire fuel cell sandwich is developed. A pseudo-homogenous model is employed to describe the electrochemical reactions in the catalyst layers. Three transport mechanisms that govern water transport through the proton exchange membrane are taken into account. Water phase change, heat generation in the catalyst layers, and thermal transport phenomena in the entire fuel cell domain are considered. The computational model is validated by experimental results and theoretical analyses, and the predicted results show satisfactory agreement with experimental and analytical ones.

Steady-state and transient numerical results of vapor, liquid water, and thermal distributions and transport in the fuel cell are presented. The influence of pressure and flow velocities on water diffusion and thermal transport are studied, and the results show that the anode water removal scheme is feasible for optimal water management as well as thermal rejection. This fuel cell model may be incorporated in a generic fuel cell power system descriptor for quantitative mass and thermal analyses and its optimal management from an overall engineering system point of view.

MATERIALS APPLICATION DEVELOPMENT, LLC

YOUR CATALYST FOR ADVANCE MATERIAL SOLUTIONS

Carbon Fiber Textiles For Fuel Cells

Author's

Paul Kennedy
Materials Application Development LLC
27 Squire Dr.
Suite # 2
North Hampton, NH 03862
603-964-4152
pkennedy@madllc.net

Jim Crawford
Crawford Associates
Rye, NH
603-661-1929
jcrawford@metrocast.net

Why Carbon Textiles Are Used In Fuel Cells

- Inert (non-reactive to Pt catalyst)
- Design Flexibility
- Negative CTE
- Low Density
- High Thermal/Electrical Conductivity
- High Strength/Stiffness

Carbon Fibers Have Low Density

CARBON FIBERS	1.76 g/cc
GRAPHITE FIBERS	2.0 – 2.2 g/cc
E GLASS	2.6 g/cc
6063 ALUMINUM	2.7 g/cc
STEEL	7.9 g/cc
COPPER	8.9 g/cc

Carbon Fibers Have High Strength

CARBON FIBERS	500 – 800 ksi
GRAPHITE FIBERS	200 – 350 ksi
E GLASS	500 ksi
6063 ALUMINUM	45 – 75 ksi
STEEL	65 – 85 ksi
COPPER	50 - 65 ksi

Carbon Fibers Have High Modulus

CARBON FIBERS	33 msi
GRAPHITE FIBERS	50 – 120 msi
E GLASS	11 msi
6063 ALUMINUM	10.6 msi
STEEL	29 msi
COPPER	17 msi

Thermal Conductivities of Materials (W/mK)

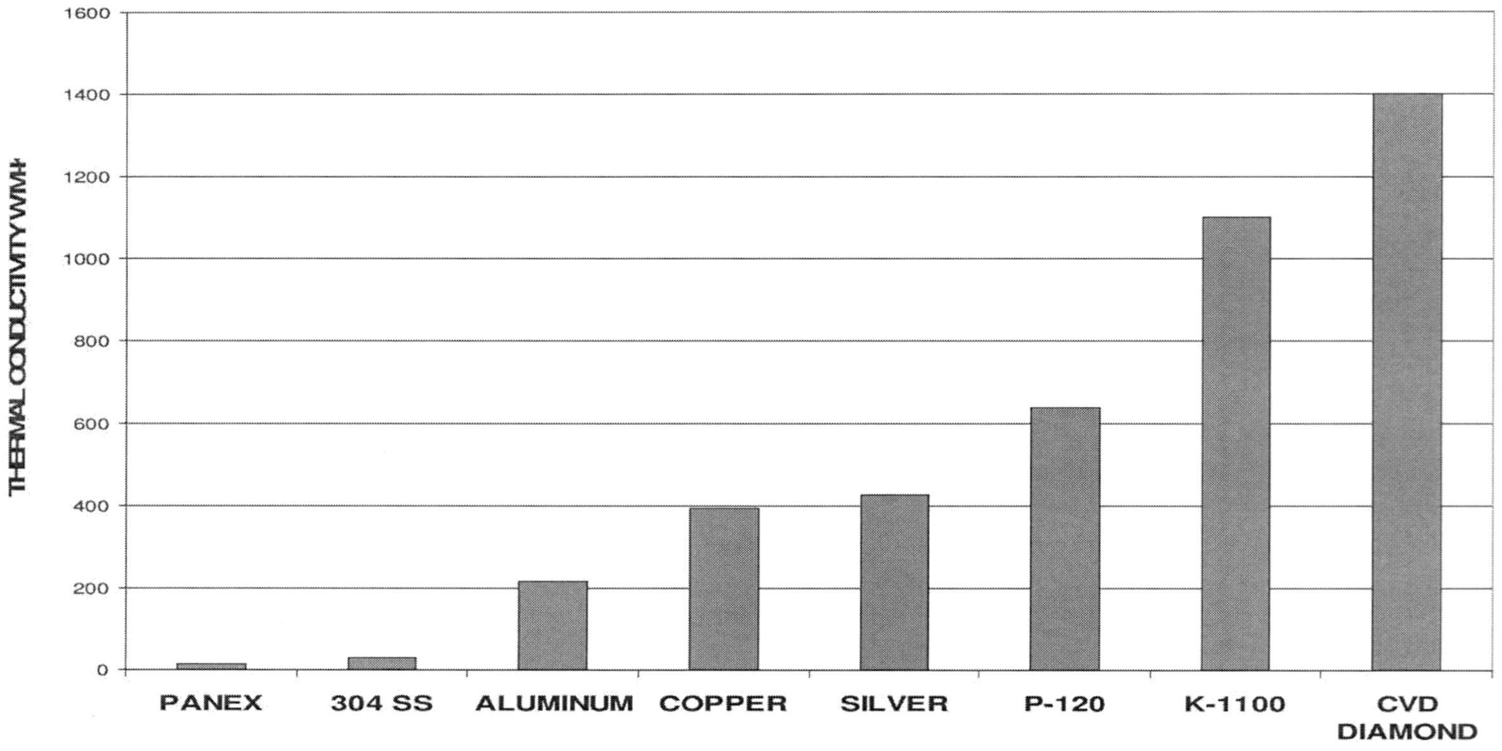

Electrical and Thermal Properties

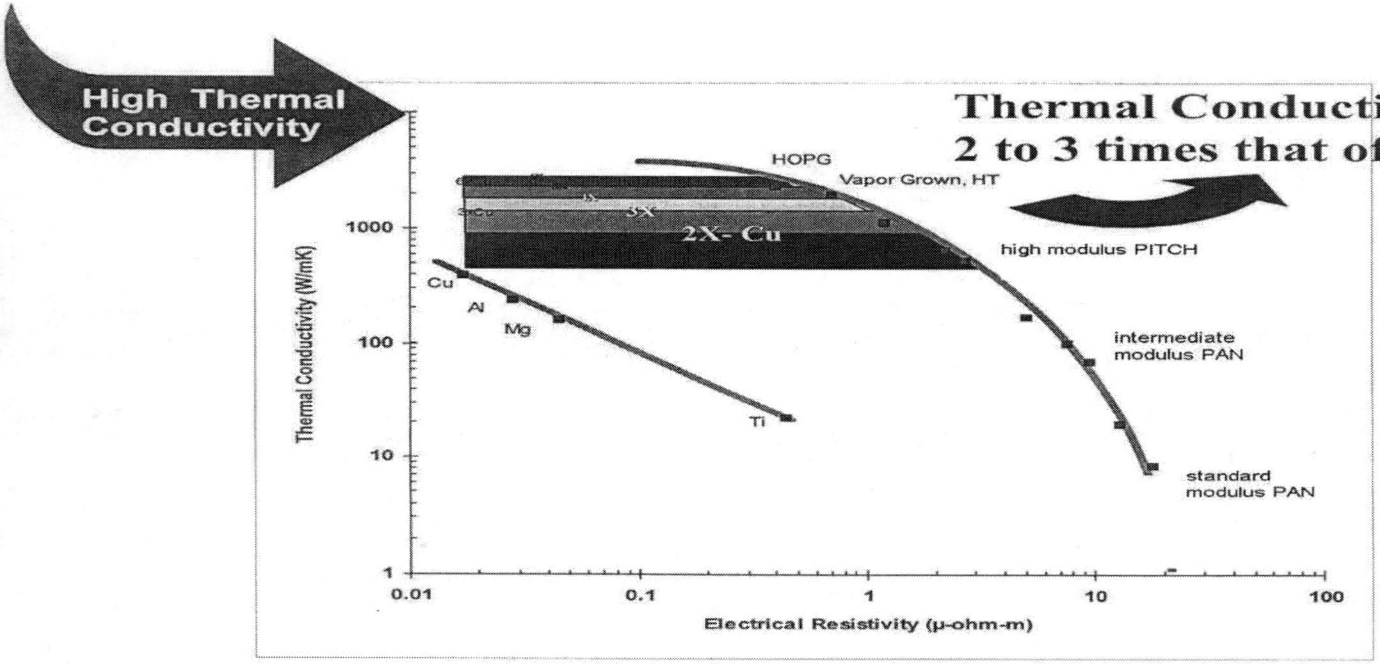

Comparison of Thermal and Electrical Behavior of Graphite

Basel Planes of HM Pitch Fibers

Structure of Graphite Crystal

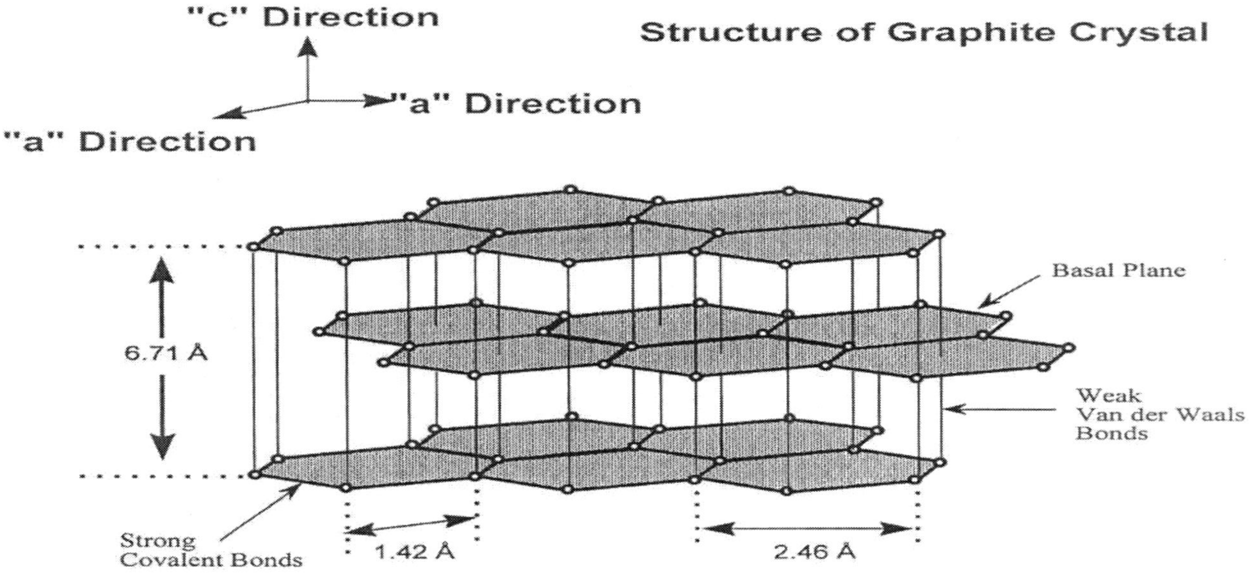

Ultra High Modulus Pitch Based Carbon Fibers

Manufacturing Variables For Carbon Textiles

- Fiber selection
- Textile conversion processes

Fiber Selection Issues

- Fiber precursor
- Filament form
- Filament count
- Yarn structure
- Fiber processing
- Fiber properties

Fiber Precursor

- Rayon (regenerated cellulose)
- Polyacrylonitrile (PAN)
- Petroleum pitch
- Phenolic

Carbon/Graphite Fiber Properties

	Tensile Strength (ksi)	Modulus (Msi)	Density (lb/in3)	Filament Diameter (micron)	Thermal Conductivity (W/mK)	Electrical Resistivit (micro ohm-m)
Pan Fibers						
Standard Modulus	500-650	30-35	0.065	7	8,14	18-15
Intermediate Modulus	650-800	42-66	.064-.067	5,7	20-40	13,11
Pitch Fibers						
Low Modulus	200-300	23-30	.069-.072	11,10	22-50	13-11
Medium Modulus	275-300	55-75	0.072	10	120-185	8.5-7
High Modulus	350	110-120	0.078	10	520-640	2.5-2.2
High TC	340-450	130-140	0.079	10	800-1000	1.5-1.1
Rayon Fibers	100	6	0.056	10	4	36-60
Phenolic Fibers	150	3	0.054	9	8	20

Filament Form

- Continuous
  - essentially infinite length
- Discontinuous (staple)
  - typically ½ to 1 ½ inches

Carbon Fibers Have Different Cross Sectional Shapes

STANDARD CARBON YARNS HAVE 1,000 TO 12,000 FIBERS (FILAMENTS) PER BUNDLE

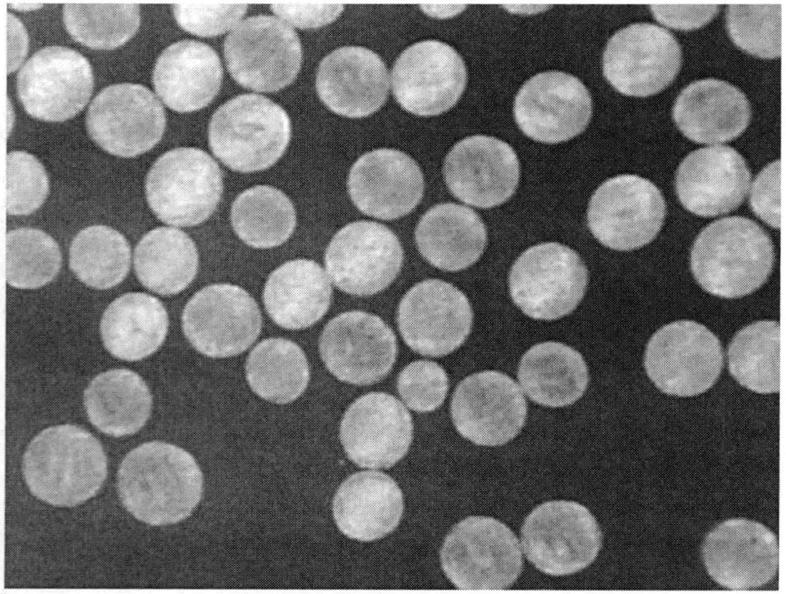

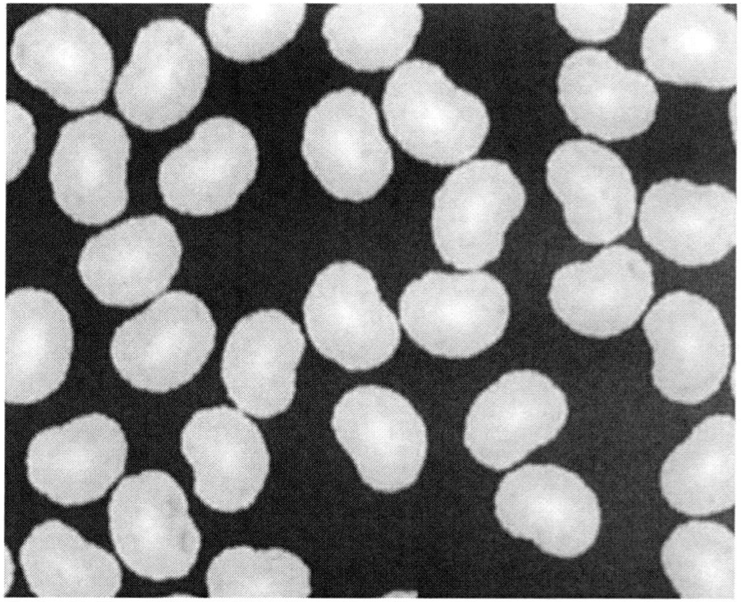

Anisotropy of Carbon Fibers

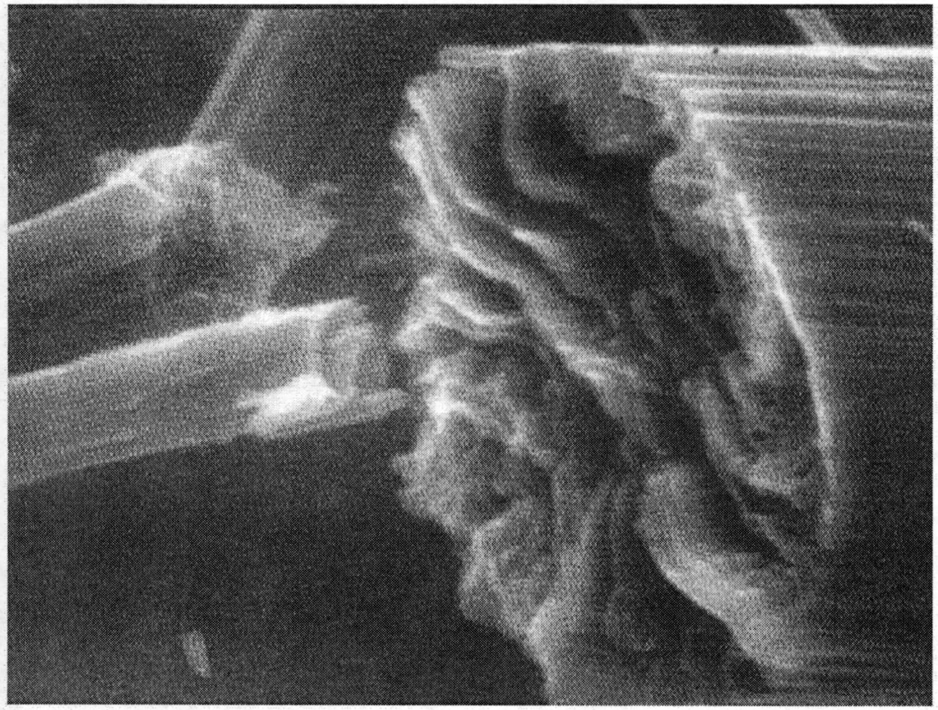

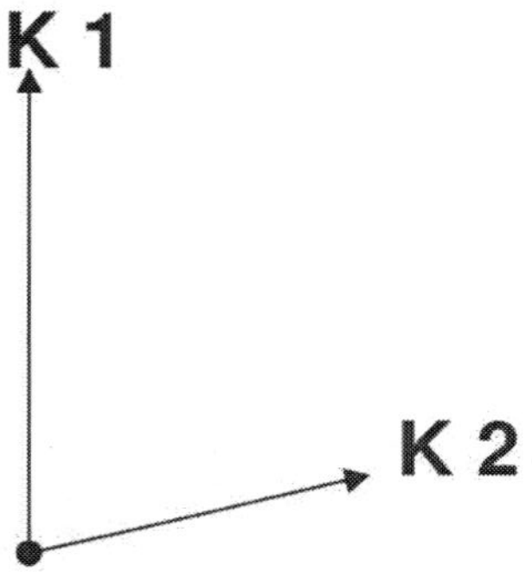

K 2 = 10 X K 1

Yarn Structure

- Continuous filament yarn forms
  - roving (no twist)
  - twisted and plied yarns
  - overwraped (served)
  - hybrids
    - continuous core/discontinuous shell

Yarn Structure (Cont'd)

- Discontinuous filament yarn forms
  - starting with continuous carbon fibers
    - stretch-broken
  - starting with staple fibers
    - carbonize a staple precursor yarn/fabric
    - twist/ply carbonized staple fibers

Fiber Processing Issues

- Process temperature
  - carbonization/graphitization
- Time at temperature
- Shrinkage
  - restrained vs. unrestrained
- Stretching

Fiber Coatings

- Sizing/coating
  - none
  - for handling
  - for bonding compatibility
- Surface treatment
  - add surface functionality
  - improve fiber/matrix bond

Typical Fiber Manufacturing Process

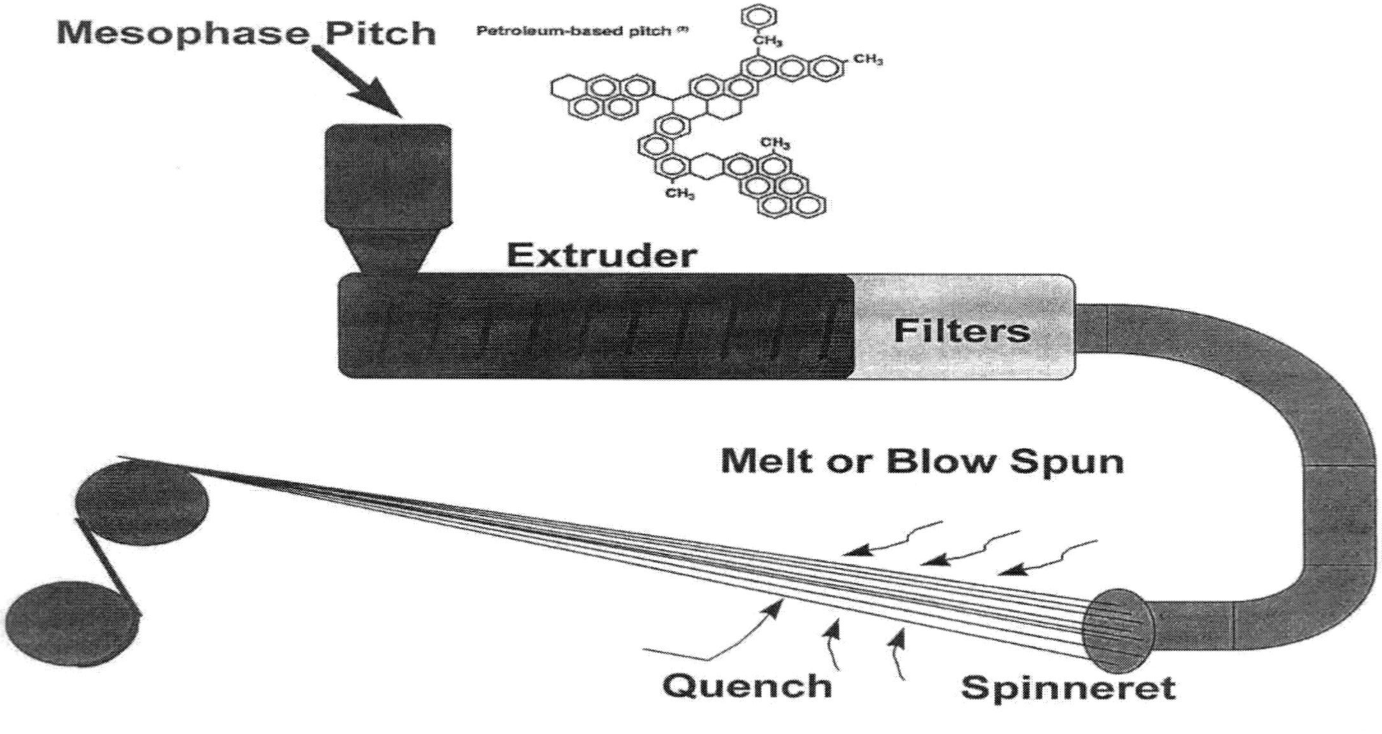

Fiber Oxidation

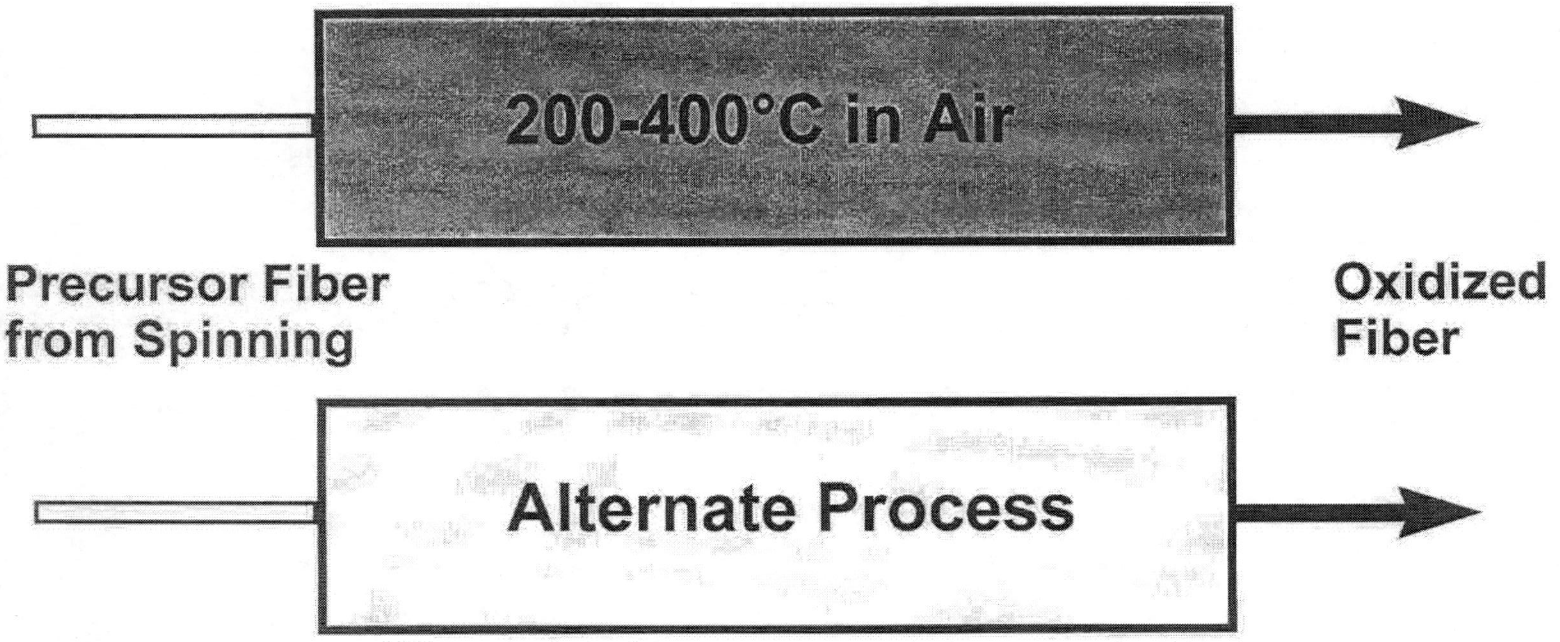

Fiber Carbonization/Graphitization

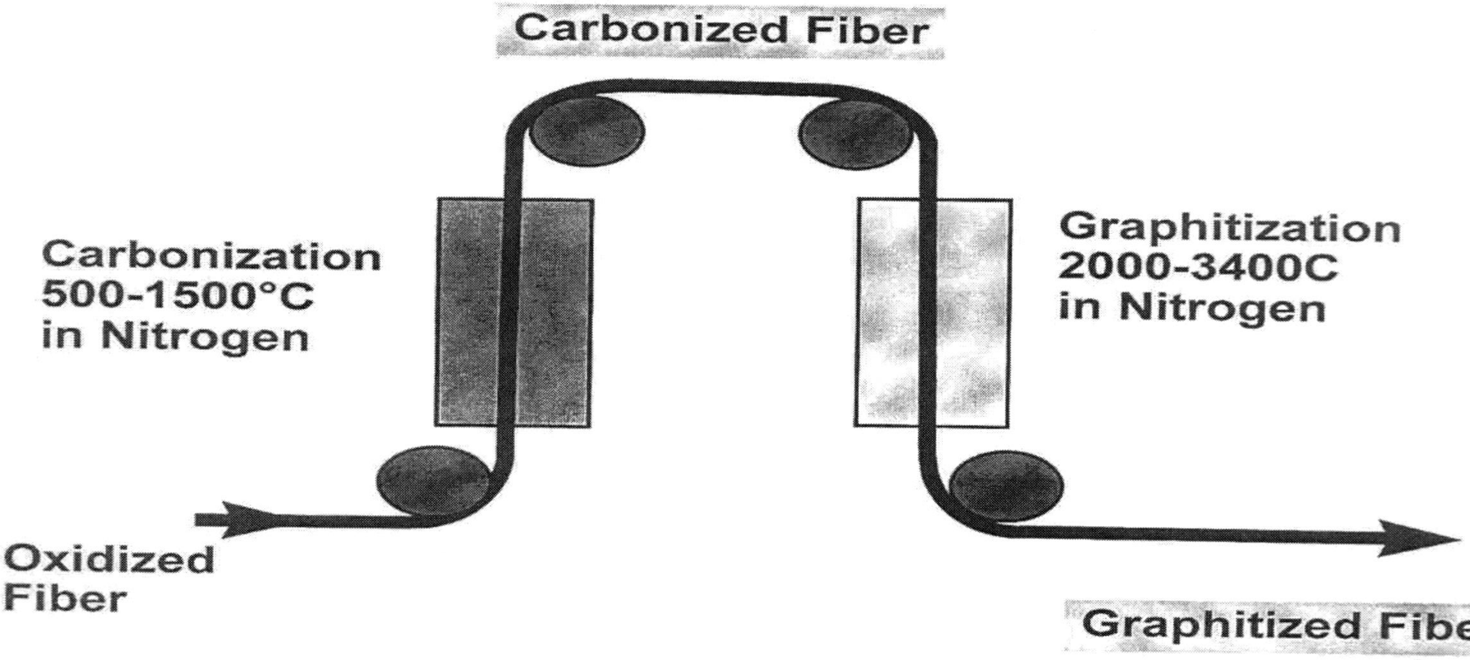

Textile Conversion Process Classification

- Non-wovens
  - needled
  - bonded
- Interlaced
  - woven
  - braided
- Interlooped
  - knitted
- Special

Needled Fibers

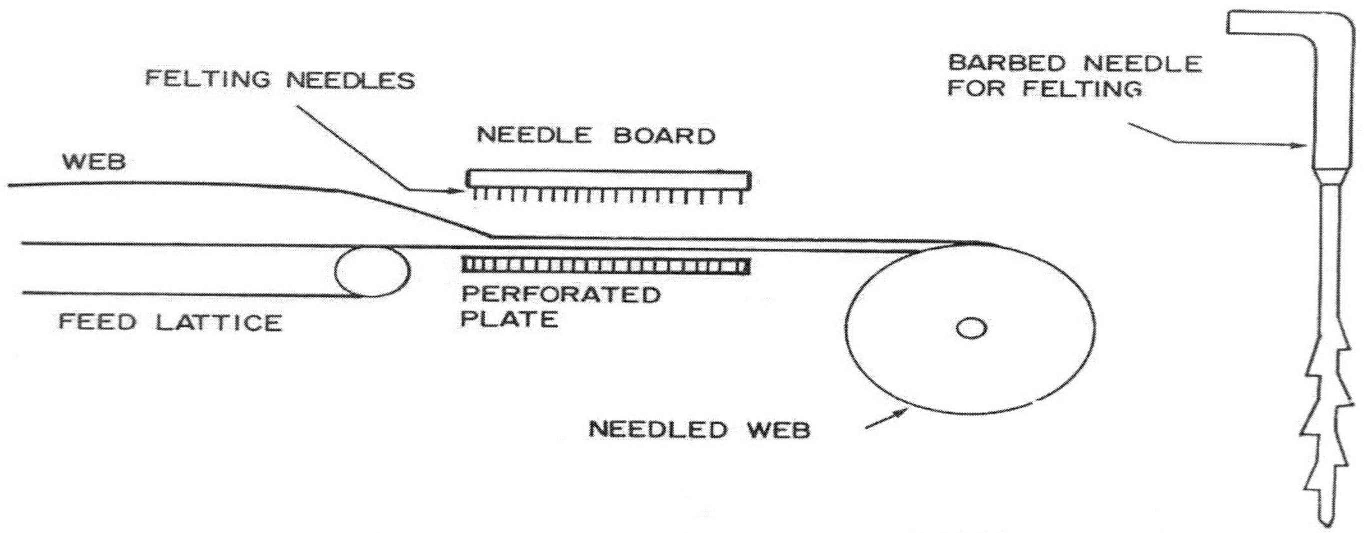

Needled Preform

Photo: Albany International

Needled Tubes

Photo: Albany International

Fabric Weave Design

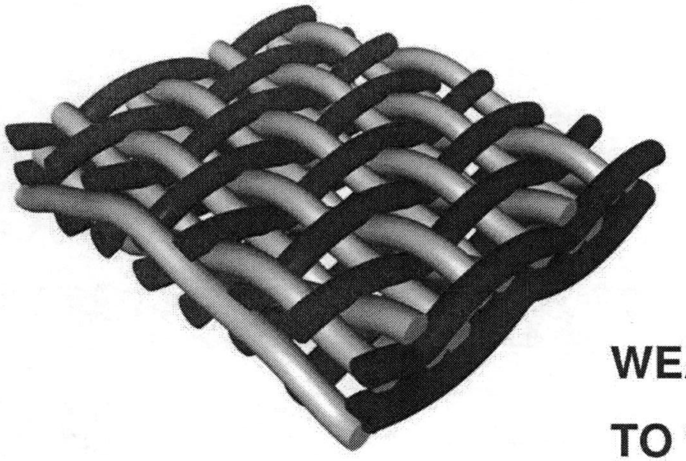

WEAVE DESIGNS CAN BE MODIFIED TO VARY FIBER CONCENTRATION AND ORIENTATION

Braided Fiber Architecture

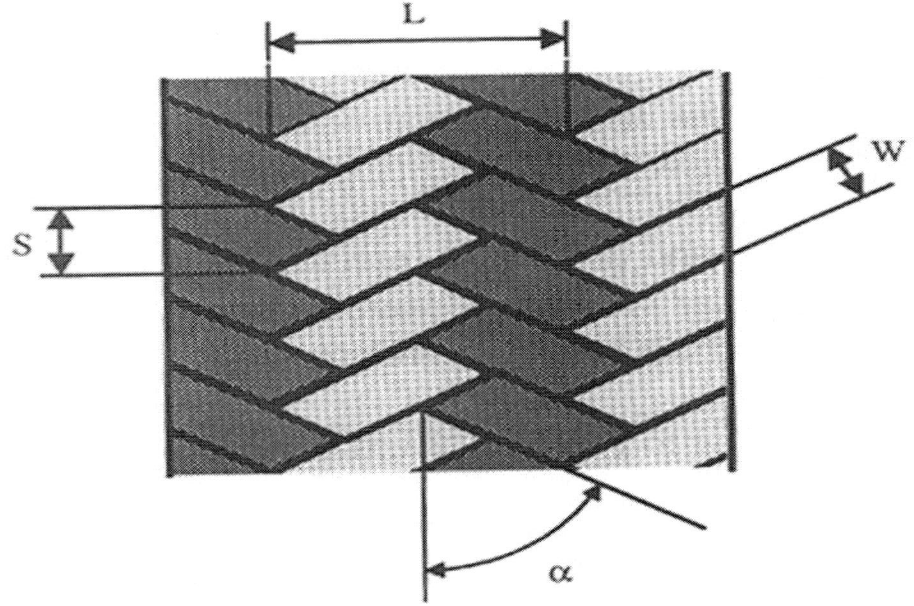

Raschel and Weft Knit Fabrics

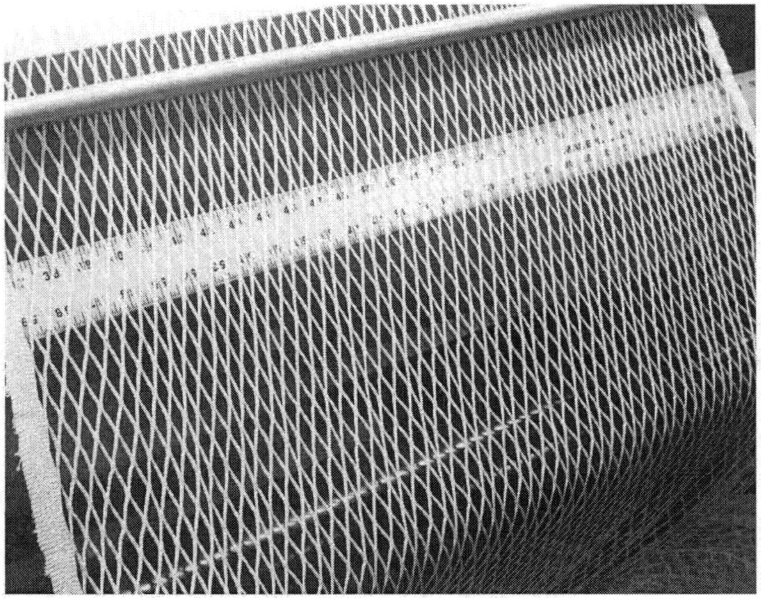

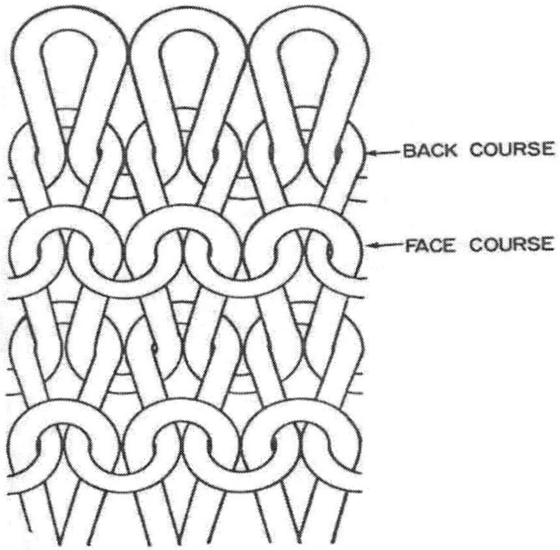

Photo: Albany International

Loop Pile Fabrics

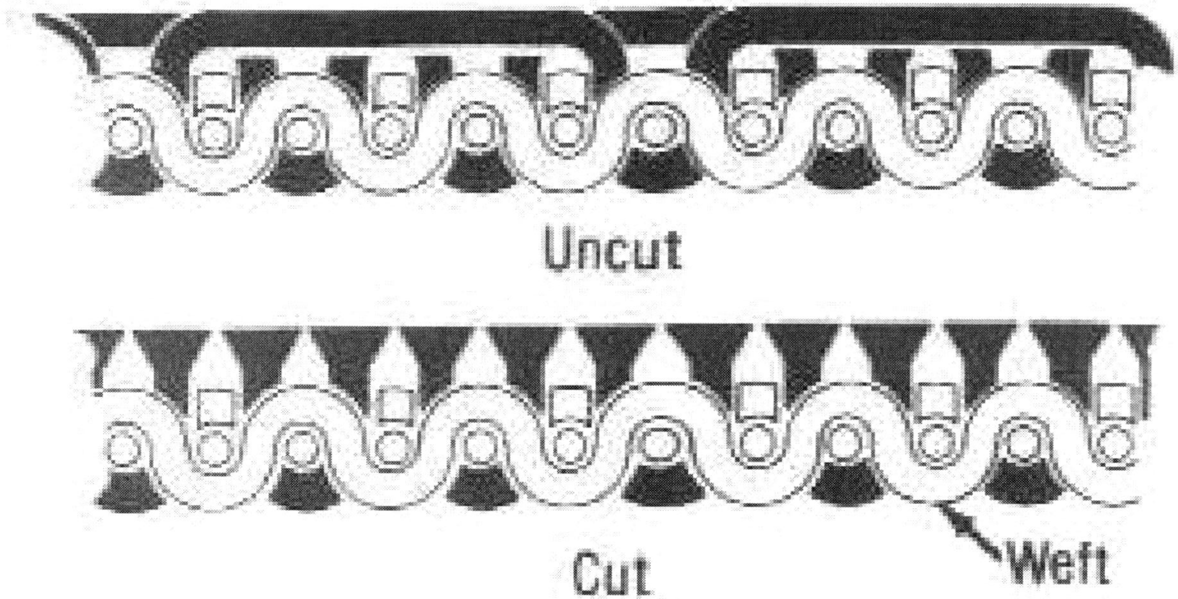

Contoured 2-D Weaving

Photo: Albany International

Helical and Shaped Weaves

Photo: Albany International

3D-Woven Preform Fabrication

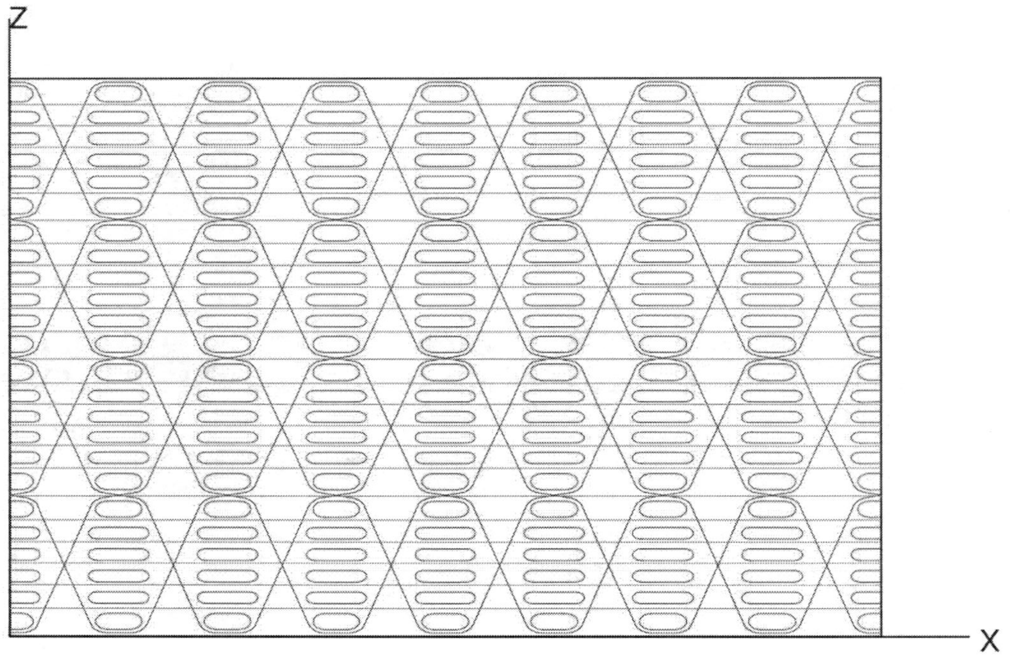

Summary/Questions

- Can improved performance be gained by:
  - High performance fibers
  - Alternate yarn forms
    - stretch broken, high modulus fibers
  - Novel weave architectures
    - alternate 2-D designs

Role of Carbon in a PEM Fuel Cell System

V K Mathur (speaker)
University of New Hampshire, Dept of Chemical Engineering
260 Kingsbury Hall
Durham, NH 03824
Phone: 603-862-1917
Fax: 603-862-3747
Email: vkm@cisunix.unh.edu

Jim Crawford
Crawford Associates
93 Wentworth Road
Rye, NH 03870
Phone: 603-661-1929
Email: jcrawford@metrocast.net

Xiaofeng Xie
Institute of Nuclear and New Energy Technology
Tsinghua University
Beijing, 100084
China
Phone: 86-10-89796077
Fax: 86-10-62771150
Email: xiexf@tsinghua.edu.cn

Carbon is a critical material in fuel cells since no other practical material has the necessary properties of electrical conductivity, corrosion resistance, surface properties and low-cost factors which could make an inexpensive fuel cell a reality. Carbon is a nonmetallic substance with a wide range of crystalline and amorphous structures. Carbon exhibits large variation in chemical stability. This paper discusses some of the salient features of carbon and its various functions in a PEM fuel cell. Table I presents some typical properties of carbon

Table I. Typical Physical Properties of Carbon

Triple Point	~4020 K and ~110 bar
Normal melting point	~3820 K
Normal boiling point	4470 ~ 5070 K
Specific gravity diamond graphite amorphous carbon	 3.515 2.266 1.8 – 1.9
C-C bond length	0.154 nm
Electrical resistivity [μΩm] carbon, petroleum coke base	 35 ~ 46

	33 ~ 66
carbon, anthracite base	58 ~ 81
carbon, lampblack base	46 ~ 66
carbon lampblack graphitized	> 600
carbon fiber	30 ~ 50
glassy carbon	8 ~ 13
graphite, petroleum coke base	~42
graphite fiber	

Fuel cells are moving past the development stage and into trials for real applications. The main components of a PEMFC power source are (i) single cell containing the porous gas diffusion electrodes (anode and cathode), the proton conductivity electrolyte, anodic and cathodic catalyst layers, and current collectors with reactant flow fields, (ii) a stack of cells in series or parallel. The carbon is used as graphite plates (Poco™) or carbon powder (Vulcan XC-72, Cabot) in many cell components. The performance of a cell/stack depends upon the graphite bipolar plate, carbon content in catalyst, carbon particle size and distribution and mode of coatings.

A PEM fuel cell consists of a positively charged electrode (anode), a negatively charged electrode (cathode), between which the electrolyte membrane is sandwiched. On one side of the electrolyte hydrogen is passed through the flow fields of the anode current collector (bipolar plate) while on the other side of the electrolyte, oxygen is channeled through the flow fields of the cathode current collector. At the anode, hydrogen reacts at the electrocatalyst sites on a gas diffusion layer (GDL) and splits into protons and electrons. The protons travel through the proton conducting membrane and react with oxygen at the cathode on the catalyst sites on another GDL and form water. The electrons at the anode are forced through the external circuit having the load, and reach the cathode, thus causing electric flow.

Bipolar Plate/Flow Field Plates

The basic form of the bipolar plate is a flat piece with carefully shaped grooves in it for gas flow. Graphite is a common material for the bipolar plate. The gas flow grooves have to be carefully machined using a slow and expensive machine. The cost of the bipolar plate is a significant part of the cost of a fuel cell system.

Electrocatalysts

Electrocatalysis may be defined as the acceleration of an electrode reaction by a substance which is not consumed in the overall reaction. The chemical steps of an electrochemical reaction can be catalyzed by a suitable heterogeneous catalyst deposited on the outer surface of a smooth electrode (membrane) or on the inner surface of a porous electrode (GDL). In PEM fuel cells that operate on hydrogen, platinum is the catalyst of choice. The action of heterogeneous catalyst is determined by the surface, the particle size and the crystal structure. Supports are necessary to increase the surface area and activity of the catalyst. Carbon powder is the most common catalyst support.

Johnson Matthey has a proprietary method for making supported catalyst; that is, attaching the minute particles of platinum to the particles of carbon black. The average size of the platinum particles is only about two nanometers. By using carbon to disperse the platinum a dramatic reduction in Pt loading has been achieved. The PTFE binds the carbon blacks particles together to form an integral porous structure, which is supported on a porous carbon paper/cloth substrate. The carbon paper GDL serves as a structured support for the electrocatalyst layer as well as acting as a current collector.

The catalyst layer remains in direct contact with the membrane. This layer consists of the catalyst and binder. The cell reactions take place on the active catalyst sites of this layer and it needs to be in close proximity to the electrolyte membrane. PEM fuel cells, being low temperature fuel cells, need to employ electrocatalysts to speed up the cell reactions. The anodic Hydrogen Oxide Reaction (HOR) is possible on a wide range of metals such as Pt and Pt alloys. They catalyze the electrochemical reaction by accelerating the splitting of H_2 molecule into H atoms, which are then ionized.

Recently, there is a major effort to deposit Pt nanoparticles on carbon nanotubes. This procedure is reported to enhance electrochemical activity, thereby improving the cell performance.

Gas Diffusion Electrode

The gas diffusion layer (GDL) is an integral part of the electrode of a PEM fuel cell. Gas diffusion layer is a slightly misleading name for is plays more than one role. It is an electrical conductor between the carbon supported catalyst and the current collector plates. It allows the flow of reactant gases H_2, and air/oxygen and the product gases through it. To improve the mass transport they need to be made porous at a loss of electrical conductivity. Often GDLs are used as the base substrates for the deposition of catalyst. GDLs are made of porous nonconductive materials such as carbon cloth or carbon paper with thickness in the range of 100-400□m. This layer typically consists of PTFE and carbon black. Carbon black imparts good electrical properties and improves the gas transport to the catalyst layer. Two types of carbon powders, namely oil furnace carbon black and acetylene black have been investigated. It is reported that acetylene black gives better cell performance than oil furnace carbon black.

Generally, GDLs are wet-proofed with a hydrophobic material like PTFE or other special material to keep it from getting wet or flooded by water. Wet proofing rejects excess liquid water so that the catalyst layer will not be flooded by water thus improving the catalyst performance. It also helps maintain the water balance in the membrane by allowing the appropriate amount of water to reach the membrane and draining the excess water.

In the dry state, the gas diffusion layers are characterized by the thickness, gas permeability and electrical conductivity. Thinner GDLs, providing less resistance to flow (i.e. high permeability) and high electrical conductivity, are generally better. Several authors have reported the use of commercial fabric GDLs like ELAT and CEER. Paper GDLs like Toray, Kureha, and SDL 10BB are also used. Toray is the most extensively used paper GDL because of the low price.

Fuelling Fuel Cells

Like batteries, fuel cells produce energy from an electrochemical reaction. An ordinary battery is limited by the amount of chemical energy that can be stored in it. But a fuel cell will operate as long as it is fed two chemical reactants, hydrogen and oxygen, from an external source.

Hydrogen fuel can be produced through thermal, electric or pyrolytic processes applied to fossil fuel, water or biomass. Nuclear-based systems can produce hydrogen from water using thermal or electrolytic processes but are not as efficient or cost effective as fossil fuels. The thermal production procedure that uses steam to produce hydrogen from natural gas or other light hydrocarbons is most common. Methane or steam reforming is typically carried out over a nickel-based catalyst that involves reacting natural gas or other light hydrocarbons with steam. Ninety-five percent of the nine million tons of hydrogen produced every year in the USA is produced via methane reforming. It can be stored in tanks or distributed from production and storage sites via trucks, railroad cars or pipelines. The use of hydrogen as an energy carrier will require development in several industrial segments, including production, delivery, storage, conversion and end use. Metal hydride systems, especially Mg-based, have high hydrogen storage capacity. These hydrides can be used to release hydrogen for fuel cell operation.

In 1989, researchers began experimenting with a revolutionary concept: using liquid methanol directly in a fuel cell as a hydrogen carrier. A solution of 3% methanol and 97% water is introduced on the anode side of the membrane. The liquid methanol spontaneously reacts with a catalyst to break the carbon hydrogen bond, with hydrogen atoms losing their electrons and becoming protons (or hydrogen ions). The protons migrate across the membrane to the cathode, while the electrons are forced around the membrane, creating an electric current. At the anode, the remaining carbon combines with oxygen from the water to produce small amounts of carbon dioxide. The direct methanol fuel cell (DMFC) faces the problem of methanol crossover and catalyst poisoning due to carbon monoxide.

In order to generate hydrogen, one can also pre-process gasoline, diesel oil or even biomass at the fuel cell power generation site.

References

1. Kordisch, K. "Fuel Cells and their Applications" VCH, Weinheim, Federal Republic of Germany, 1996.

2. Larminie, J, Dick Andrew, Fuel Cell Systems Explained, John Wiley and Sons, New York, 2002.

3. Koppula, K., M.S. Thesis, University of New Hampshire, Durham, NH, 2004.

4. Mathur, V.K., Xie, X., Hamblin, T., Materials: Key to Hydrogen Economy, Proceedings, Spring AIChE Meeting, April 2004.

120d The Role of Thermodynamics and Rate Processes with Respect to Alternative Fuels

Michael Neuman, Stuart W Churchill

When widespread agreement to protect and clean up the environment emerged in the 60's the most commonly proposed model of action with respect to the chemical industry was simply to reduce or prevent end-of-the-pipe pollution. A much broader concept has since evolved that incorporates the concept of sustainability. Sustainability is essential in the long run because all resources are subject to the first and second laws of thermodynamics. The first law reveals that all resources are finite and that their exploitation invokes inexorable tradeoffs. The second law reveals limitations and consequences of the possible tradeoffs.

The objective here is to consider sustainability in terms of these limitations and consequences. Both laws require the careful choice of an envelope or system, for example, a single apparatus, a single processing unit, an entire plant, a geographical region, the entire earth and its atmosphere, or the universe.

The second law tells us that sustainability is not possible in terms of the total availability (or exergy) but instead represents a goal to be approached within some particular envelope at the expense of changes outside the envelope. The first law leads us to express the conservation of energy and quantities such as chemicals in terms of accumulation and depletion within the envelope and rates of transport through the chosen envelope.

Thermodynamics thus indicates the limits of what can be done within any framework of space. On the other hand, the rate processes such as fluid flow, heat transfer, mass transfer, and chemical reactions determine the time required to carry out transformations, which is of equal importance.

Social, political, and economic considerations guide our choices among those that are possible thermodynamically. Unfortunately, these choices often ignore what must occur outside the envelope to accomplish a desirable end within the envelope.

Hydrogen Production by Dissociation of Water through a Novel Membrane

R.J.Boersma, J.A.Shindle
CellTech Power Inc, 131 Flanders Road, 01581 Westborough, MA.

Abstract

This paper addresses a novel method of hydrogen generation, directly from carbonaceous fuels by utilizing a membrane with mixed ionic - electronic conducting properties. Unlike conventional methods where hydrogen is separated from a synthesis gas stream, this technique generates a pure hydrogen stream through the electrochemical dissociation of water. The process exploits the difference in partial oxygen pressure between a rich fuel stream and a lean fuel stream. The rich fuel can be a gasified liquid fuel or coal or any other fuel stream and the lean fuel can be steam. The oxygen partial pressure differential represents an electrochemical voltage differential which is the driving force for the dissociation of water on the lean side into hydrogen and oxygen. The oxygen, in the form of oxygen ions, is driven through the membrane where it oxidizes the rich fuel stream. The resulting wet hydrogen stream can then be used as a carbon and sulfur-free fuel source, ideally suited for fuel cells. This paper reports on an experimental membrane consisting of a mixture of Yttria doped $SrTiO_3$ mixed with Yttria stabilized Zirconia. The first phase is an electronic conductor, the second phase is an ionic conductor. Reported is the dissociation of water in the temperature range of 800-1000°C which was found to be 0.4-1.8 $cm^3/(cm^2\ min)$.

Introduction

Conventional methods for the separation of hydrogen from carbonaceous fuels typically require the following steps: 1) a clean-up step, 2) reaction to a syngas stream, 3) a shift reaction and 4) a hydrogen separation step. In the case of liquid and solid fuels, the process will be preceded by a gasification step. In the clean-up step particulates and contaminants are removed. The syngas is obtained by reaction of a fuel with steam, air or pure oxygen to create a mixture of hydrogen, carbon monoxide, carbon dioxide, water and lower hydrocarbons. The clean syngas stream is subsequently reacted in a cascade of reactors to form as much hydrogen as possible through the water-gas shift reaction. A subsequent separation step is required to remove the CO_2 which in the conventional process, is typically done by pressure swing adsorption (PSA), a process which is energy intensive.

The energy demand and process complexity are among the reasons that more advanced separation methods are being developed. However, with many of the advanced technologies that are being developed, the number of process steps is still significant.

This paper addresses a Mixed oxygen Ion– Electron Conducting (MIEC) ceramic membrane that dissociates water electrochemically to produce hydrogen by using the enthalpy from a carbonaceous fuel stream, all in a single process step. This is called chemically driven electrolysis. MIEC materials have received attention because of their potential to separate oxygen from air [1] and to feed pure oxygen into a methane stream to accomplish its partial oxidation with the objective of obtaining a syngas that is not diluted with atmospheric nitrogen [2]. These ceramics are capable of conducting oxygen ions and electrons, properties common to many metal oxides. Particularly, materials with the perovskite structure have been studied for these properties. Conductivities depend on the element combinations, their abundance in the structure, doping materials, gas partial pressures, temperature and sintering conditions and can vary by many orders of magnitude. The properties of many MIEC's have been researched as a function of these variables and the intent of this work was to identify among these, a material that would meet the operating requirements for a water dissociating membrane as would be encountered in a real life reactor.

The principle of chemically driven electrolysis process was first described in 1977, by Browall et al [3]. In earlier work, researchers found that by applying an external voltage to an electrochemical cell, oxygen could be driven into a carbonaceous fuel such as methane, whereby water dissociated into hydrogen. The advantage of this chemically assisted electrolysis method is that the electrical energy required is only a fraction of the electrical energy required to dissociate water as in the conventional electrolysis process. When the voltage leads of the chemically assisted process are shorted, then the electrolysis becomes chemically driven. However, since the voltage being shorted is low, only small currents will arise if the resistance of the electrical circuit is high. This calls for an approach with the shortest possible electronic current path as found in MIEC materials. Of course both ionic and electronic conductivity need to be as high as possible and if one is insufficient then the process will not work effectively as was found by Naito et al [4]. Other researchers have successfully demonstrated that it is indeed possible to dissociate water with a MIEC membrane. For instance, Balachandran et al [5] describe a mixture of a metal and YSZ to obtain a two phase MIEC, where the metal acts as the electronically conducting phase. Gopalan [6] describes single phase MIEC materials based a $LaSrCoFeO_3$ system, though identifying a single phase material that has desirable electronic and ionic conductivity throughout the required oxygen partial pressure range and that is also stable in a sulfur bearing gas may prove to be difficult. Related to this work is a two phase MIEC described by Mazanec [2], who uses the membrane to accomplish partial oxidation of methane. Described are results obtained with an ionic phase of 8YSZ mixed with various electronic phases. The best results were obtained with an electronic conductor made of Pr-Zr doped Indium.

Another approach to produce hydrogen using MIEC membranes is through proton conducting materials. In this case, water is not electrochemically dissociated, but hydrogen travels as protons through the membrane. To maintain overall charge neutrality protons and electrons must flow in the same direction. Protonic MIEC's based on $BaCeO_3$ have been researched by Balachandran et al [7], however it was concluded that the protonic conductivity was insufficient for use as a hydrogen separation membrane.

Theoretical background

As Figure 1 shows, oxygen ions created on the cathode side of the MIEC membrane can travel through the ionic phase of the material and release its charge upon leaving the membrane. The electrons, flow back through the electronic phase of the membrane, thereby completing the electric circuit. An oxygen partial pressure differential across the membrane is required for the process to proceed which due to the ionization of the oxygen gives rise to an electrical field which forms the driving force for the passage. The principle is equivalent to a fuel cell, the difference being that the electronic conductivity of the membrane creates a low-resistance internal short circuit.

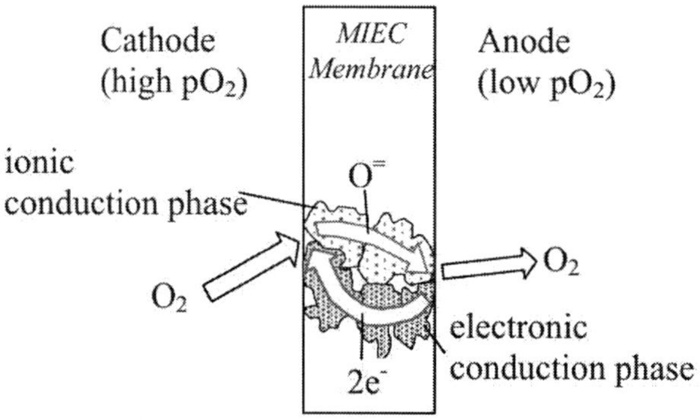

Figure 1. *Schematic representation of the process taking place inside a mixed ionic electronic conductor (MIEC)*

The magnitude of the electrical field is calculated from the Nernst equation:

$$V = \frac{-\Delta G^{\theta}{}_{T}}{2F} - \frac{RT}{4F}\ln\left(\frac{pO_{2,a}}{pO_{2,c}}\right) \qquad \text{Eq.1}$$

Here V is the Nernst voltage, $\Delta G^{\theta}{}_{T}$ is the free enthalpy change of the net reaction at unity concentrations, T is the absolute temperature, $pO_{2,a}$ and $pO_{2,c}$ are the oxygen partial pressures at anode and cathode respectively, R is the universal gas constant and F is the Faraday constant. Contrary to the situation with a fuel cell the Nernst voltage cannot be measured on the membrane itself since the electrodes are always shorted. Nevertheless, it is a measure for the chemical potential difference between the two sides.

The reaction equations at the electrodes are:

$$\begin{aligned} &\text{Cathode:} \quad \tfrac{1}{2}O_2 + 2e \rightarrow O^{2-} \\ &\text{Anode:} \quad O^{2-} \rightarrow \tfrac{1}{2}O_2 + 2e \end{aligned} \qquad \text{Eq.2}$$

Consider, for instance, the case where there is a CO/CO_2 mixture on one side of the membrane and an H_2/H_2O mixture on the other side, both in ratios as shown in Table 1. Also shown are the O_2 equilibrium pressures for the both mixtures at 800°C and the associated Nernst voltage as per equation 1.Since the net reaction from equation 2 does not constitute a chemical change $\Delta G^{\theta}_{T} = 0$ in equation 1.

Table 1. *Nernst voltage resulting from two gas compositions at 800°C and standard pressure.*

anode	$pH_2/pH_2O = 0.1$	$pO_2 = 3.57E\text{-}17$	V=0.203 volt
cathode	$pCO/pCO_2 = 8$	$pO_2 = 6.10E\text{-}21$	

The reaction mechanism has been described in terms of partial oxygen pressures, but it is equally valid to say that effectively water dissociates at the membrane into an oxygen ion and a hydrogen molecule. The ion travels through the membrane and oxidizes carbon monoxide to carbon dioxide, thus:

$$\text{Cathode:}\quad H_2O + 2e \rightarrow H_2 + O^{2-}$$

$$\text{Anode:}\quad CO + O^{2-} \rightarrow CO_2 + 2e \qquad \text{Eq.3}$$

Summation of these reactions gives:

$$H_2O + CO \Leftrightarrow H_2 + CO_2 \qquad \text{Eq.4}$$

This is the well known water – gas shift reaction and therefore to predict the temperature and pressure dependence of the process the thermodynamics of the water-gas shift reaction needs to be considered.

Since the circuit is shorted by the electron flow through the electronic phase an electrical current, I, according to Ohm's law, can be expected:

$$I = \frac{V}{R_m} \qquad \text{Eq.5}$$

The total resistance, R_m, can be broken down into the following main components:

- A polarization resistance on the cathode due to the charge transfer, R_c
- An ohmic resistance resulting from the ionic transport through the membrane, R_i
- A polarization resistance on the anode due to the charge transfer, R_a
- A electronic resistance that short circuits the cell, R_e

Therefore:

$$R_m = R_c + R_i + R_a + R_e \qquad \text{Eq.6}$$

The ionic resistance is likely to form a substantial proportion of the total resistance. It can be minimized by reducing the thickness to the minimum that is practically and reliably achievable, while remaining fully dense. The electronic resistance, R_e, is, as will be seen, because of the materials choice, negligible relative to the total resistance, R_m. The polarization resistances depend on the surface characteristics, the strategy to minimize these is to increase the reaction contact area by using high specific area powders with catalytic properties.

Materials

In this paper, we report on a two phase MIEC material, having a ceramic ionic phase and a ceramic electronic phase. It was perceived that the generally similar properties of ceramics would lead to a reasonably simple manufacturing process, while being able to select the most appropriate individual phases. Stable properties throughout the partial oxygen pressure range from about 10^{-15} to 10^{-25} bar are required. In addition, mechanical ruggedness is a consideration, as well as thermal cycling capability. Furthermore, sulfur tolerance and surviving an inadvertent exposure to a high oxygen concentration are important. A potential drawback of the two phase approach is the potential of chemical and physical interference of the individual phases which limits the possibilities.

The following materials were evaluated:

- 8YSZ (8mol%Y_2O_3 stabilized ZrO_2) and 10ScSZ (10mol%Sc_2O_3 stabilized ZrO_2), both ionic conductors;
- CGO (Ce doped Gd_2O_3); mixed conductor at low pO_2, ionic at high pO_2;
- LSCrF ($La_{0.2}Sr_{0.8}Cr_{0.2}Fe_{0.8}O_3$); ionic conductor at low pO_2, mixed at high pO_2;
- YST ($Sr_{0.88}Y_{0.08}TiO_3$ Yttria doped Strontium Titanate); electronic conductor at $pO_2<10^{-13}$;
- LCC ($La_{0.8}Ca_{0.2}CrO_3$, Calcia doped Lanthanum Chromite) electronic conductor at $pO_2<0.2$;

The materials 8YSZ, 10ScSZ and CGO are well known materials from the fuel cell industry. 8YSZ and 10ScSZ are oxygen ion conductors with negligible electronic properties. At 800°C the ionic conductivity of 10ScSZ is more than 4 times higher than that of 8YSZ. These materials are used for the electrolyte in SOFC's. The material CGO is added to the anode and cathode of low temperature SOFC's with the purpose of improving the ionic conductivity. It is also being used as barrier layer between cathode and electrolyte to prevent a chemical reaction between the two. At 800°C the ionic conductivity of CGO is over 6 times that of 8YSZ, though mechanical properties and chemical stability of CGO are in general inferior to those of the Zirconias.

LSCrF is being studied as a material to separate oxygen from air [8]. Its ionic conductivity is almost 10 times the conductivity of 8YSZ at 800°C. When LSCrF was sintered onto a porous 8YSZ substrate the LSCrF turned out to be highly reactive; it also reacted with the alumina kiln furniture. Therefore its application on a porous substrate was considered too problematic, for which reason it was abandoned.

The YST is a material that has been extensively studied by Hui et al [9]. It was developed as an anode material for the SOFC. The electronic conductivity is over 60S/cm at 800°C which is sufficient for our purpose. When going from a reducing atmosphere to an oxidizing atmosphere the lattice parameter changes only slightly, indicating that the material is stable to this type of cycling. An alternative to YST could be La doped $SrTiO_3$. Conductivities at doping levels up to 40% La were measured by Marina et al [10] and Pudmich et al, [11]. Conductivities of well over 100 S/cm were measured below a partial oxygen pressure of 10^{-13} bar; above this value the conductivity reportedly drops off to about 50 S/cm. At even higher partial oxygen pressure the conductivity becomes negligible and the material was reported to crack.

Mixtures of LCC and Zirconias have been studied and used extensively at CellTech Power in their proprietary liquid anode fuel cell technology. LCC would make an excellent candidate was it not for the fact the Cr is mobile and interferes with the ionic conductivity of the Zirconia phase.

The combination of the Zirconias with YST appears favorable. The Zirconias have desirable chemical and mechanical properties and are well studied. The electronic conductor YST is a relatively new material which fact brings along some uncertainty. The fact that the material is stable to reduction - oxidation cycles is a pre from a real life operating perspective.

The thermal expansion coefficient of YST is, as shown in Figure 2, ranging from $11\text{-}12 \cdot 10^{-6}\,K^{-1}$; values highly compatible with those of YSZ. An XRD analysis of an 8YSZ-YST sample sintered in air at 1500°C showed peaks of 8YSZ and $SrTiO_3$ only, no additional peaks were showing within the detection limits of the apparatus (Figure 3). $SrTiO_3$ was used as a reference since the Yttria doped version was not present in the materials database which also explains a small shift of some of the peaks relative to the reference. Therefore it can be preliminary concluded that there are no new phases being formed and that reactivity between the phases may not be an issue. Essentially the same conclusion was drawn by He et al [12]. It is conceivable, however, that a new phase is being formed at the grain boundaries that cannot be detected by the XRD analysis. A more definite answer can be provided from a Scanning Tunneling Electron Micrograph (STEM) analysis.

La doped $SrTiO_3$, mixed with Ce-La oxides, has been studied as a sulphur tolerant electrode for fuel cells by Marina and Pederson [13] and more results were recently presented at a peer review meeting of the SECA Core Technology Program [14]. In the latter a 30% drop in current density at an H_2S level of 950 ppm was reported, starting from a current density of around 0.65 A/cm^2. The fact that the performance is stable is significant since exposure to less than 10 ppm H_2S is detrimental for a state-of-the-art Ni anode. The polarization of the unexposed $SrTiO_3$ anode appears to be relatively high, at least 2-3 times higher than that of a Ni anode. The thermal expansion of these electrodes is reportedly similar to that of YSZ and therefore YST.

Potentially these electrode formulations can be used to catalyze the anode reaction (Equation 2) to oxidize practically any sulfur containing carbonaceous fuel and thereby reducing water on the opposite electrode. Assuming that the materials are chemically compatible, a picture emerges of a system capable of performing the intended chemically

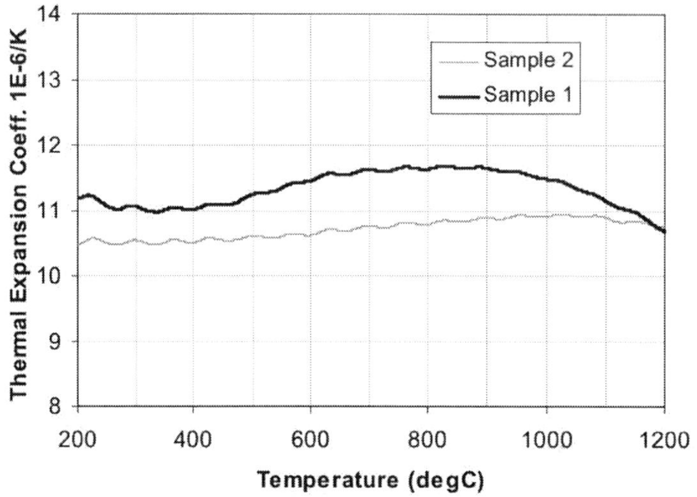

driven electrolysis on sulphur bearing fuels

Figure 2. *Thermal expansion coefficient relative to original size of two YST samples. Sample2 is as obtained from the sintering furnace, sample 1 has been reduced at 1200°C in a 95/5 Ar/H_2 atmosphere and was re-oxidized during dilatometry. In both cases thermal expansion is similar to YSZ, indicating that a YSZ-YST system may be stable in oxidation-reduction cycles*

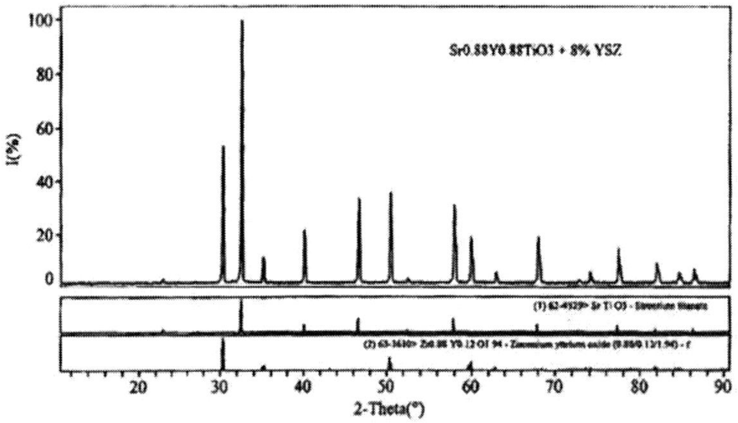

Figure 3. *XRD plot of an 8YSZ- YST sample sintered at 1500°C. Bottom two plots are from database samples from 8YSZ and $SrTiO_3$, individually. The top plot is a superposition of the database plots, which leads to the conclusion that within the detection limits no new phases have been formed.*

By using separate electronic and ionic phases in approximately equal volume fractions, one can expect the electronic and ionic conductivity of either to be less than half that of the bulk and more likely one third of the bulk conductivity which is due to the tortuosity of the path.

Interpretation of results presented by Mazanec et al [2] confirms this. The bulk conductivity of YST is reported at 64 S/cm at 800°C and $pO_2 = 10^{-19}$ bar [9] which is significantly higher than the ionic conductivities of 8YSZ (0.02 S/cm bulk) and 10ScSZ (0.094 S/cm bulk). The limiting factor will therefore come from the ionic phase. The area specific resistance of a 10 μm membrane works out to approximately $150m\Omega.cm^2$ at 800°C. This will be roughly in the same order of magnitude as the polarizations of both electrodes together, leading to an expected yield of approximately 5 $cm^3/cm^2/min$. It may be possible to increase the yield by raising the temperature; alternatively the Yttria can be replaced by Scandia in the ionic phase. It is also possible to increase the ionic phase volume fraction, though the limit will lie around 70%, since above this value the electronic phase may lose its continuity.

To enhance the charge transfer reactions electrocatalytic surfaces need to be applied on both sides of the membrane. To achieve the lowest resistance a thin membrane layer is required. For mechanical support a porous substrate, such as YSZ or YSZ-YST is used. Thus, the substrate is coated with a first electrocatalytic layer, then the membrane is applied, followed by a second electrocatalytic layer. Since the rich fuel may carry particulates with it (soot), it makes sense to have the anode as the second layer, on the exterior of the membrane layer, so that it cannot foul the porous substrate. The obvious candidate for this layer would be the doped La-Ce $SrTiO_3$ material discussed above. For the first catalytic layer a candidate material is a highly active, porous formulation of 8YSZ – YST and since the YST was originally developed as the electronically conducting phase for SOFC anodes this seems an obvious choice [9]. There is however little known about the catalytic properties of the YST and it may require further doping to improve the activity [15, 16]. However, more conventional H2-steam electrodes, such as Ni-YSZ or Ni-CGO may prove to work well for the cathode.

Experimental

For reason of availability, a tubular porous NiO-8YSZ substrate was used to test the concept. The substrate was prepared by mixing NiO, 8YSZ and carbon with binders and plasticizers to obtain a clay-like substance that was extruded through a 7.5 ID X 9 OD mm dye. NiO and 8YSZ were mixed in a ratio that was calculated to give a Ni to 8YSZ volume ratio of 40 to 60. Carbon was added to the extent that a porosity of 40% was obtained. The formulation was optimized to meet fuel cell requirements and is not entirely relevant for the water dissociation. The extrudate was dried and sintered at 1150°C, yielding a partially sintered (bisque fired) tube. A first electrocatalytic layer was applied by dipping the tube in a colloidal suspension of Ni-CGO followed by drying and heating to 1000°C. Subsequently the tube was dipped in a colloidal suspension containing YST and 8YSZ (40/60 v/v), followed by sintering at 1450°C for 2 hours. By the appropriate choice of the bisque firing temperature the shrinkages of the bisque fired tube and the YST - 8YSZ coating, are equal. Consequently the membrane on the substrate is dense, crack free and not delaminated. After this a CGO layer was applied which layer was sintered at 1000°C The Ni-CGO and CGO electrodes are well known from fuel cells and were chosen to avoid uncertainty arising from the electrodes discussed in the materials section. The resulting tube had a diameter of 7 mm, a wall thickness of 1 mm and had a coated length of 60 mm.

A steam-H_2 mixture was fed through the inside of the tube at flow rates of respectively 1000 cm^3/min and 17 cm^3/min. In a counter-flow orientation a CO/CO_2 (80/20) mixture was fed over the outside of the tube at a rate of 250 cm^3/min (Figure 4). After passing the tube the CO/CO2 gas was analyzed in a gas chromatograph. The analyzed CO/CO_2 ratios at various temperatures are shown in Figure 4. Production rates of up to 20 cm^3/min of H2 were calculated from the analysis. Unfortunately, the seal holding the tube developed a leak which may have interfered with the measurement, due to the CO/CO_2 mixture shifting in the presence of the steam/H_2 leak. A correction was applied and the method will be discussed below. In Figure 5 corrected values are shown. On a number of occasions the flow exiting from the steam/H_2 side was measured with a bubble flow meter and was found to be in reasonable agreement with the corrected GC data (Table 2).

After testing the tube was sectioned, a SEM photograph is shown in Figure 6. The dense layer is clearly showing; the dark and light areas are YST and 8YSZ, respectively. Good connectivity is likely in this layer. A problem appears to be that the first active layer has penetrated and blocked the porous substrate. Initially it was thought that a reaction between substrate and first active layer had taken place, though as it turned out the particles in the active layer were so small that they could easily penetrate relatively deep, in some cases up to 100 μm, into the substrate. This penetration partially blocked the porous structure, requiring the ionic species to travel much longer distances than intended, leading to a lower performance than expected.

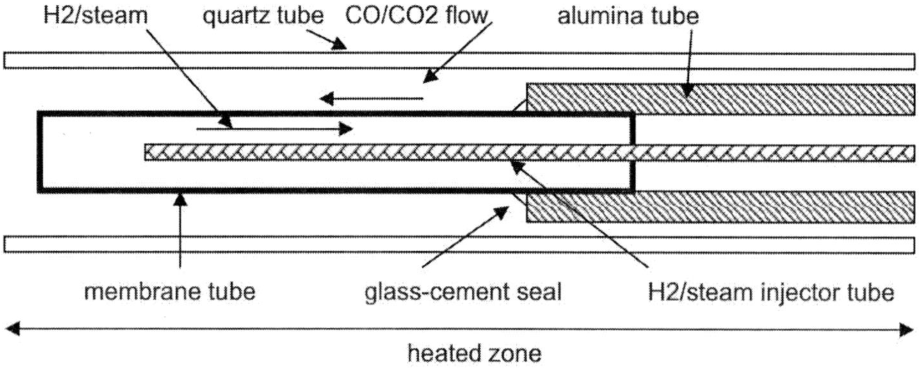

Figure 4. *Layout of the process in the experiment.*

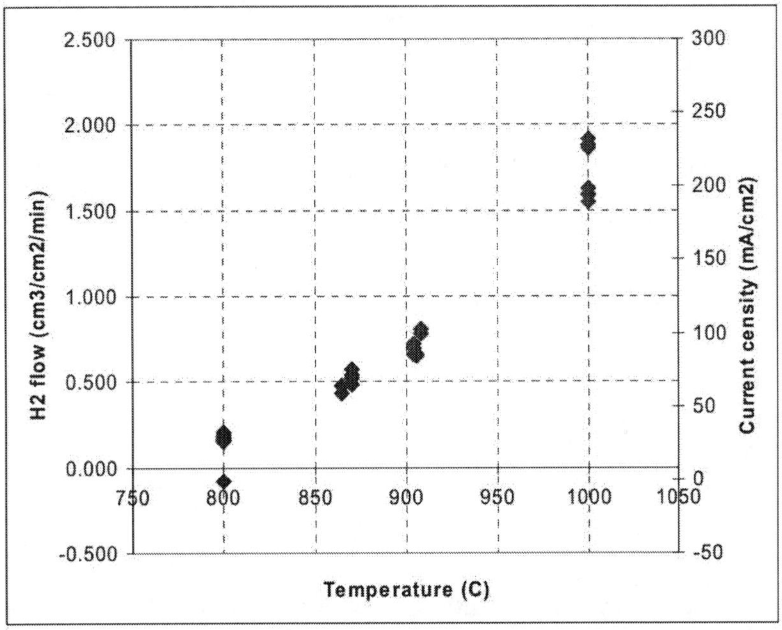

Figure 5*. Hydrogen flow, expressed per unit of cell area, calculated from gas chromatography data, as a function of temperature. Most data points were obtained at 17cm^3/min of H_2 and 1000cm^3/min of steam on the cathode side and 200 cm^3/min CO and 50 cm^3/min CO_2 on the anode side. Data points showing higher yields were obtained at 7cm^3/min of H_2 and 250 cm^3/min CO.*

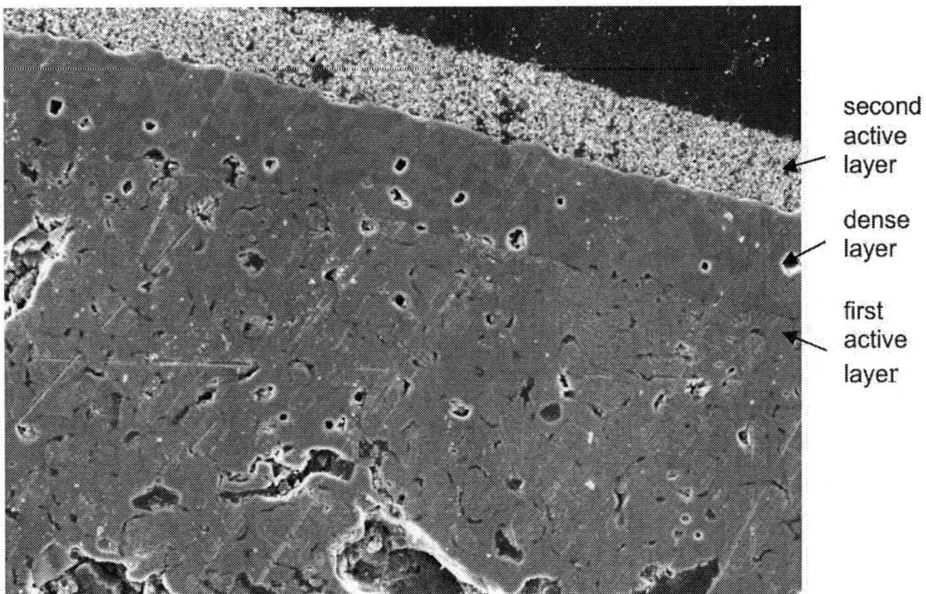

Figure 6*. SEM photograph of a cross section of the tested tube*

Table 2. *Volume flows calculated from GC data, based on CO/CO_2 composition change, compared to the flow measured at the outlet on the hydrogen side.*

Temperature	908	1000	°C
Generated H2 flow calculated from GC data	9.4	23.8	cm^3/min
Flow change on H2 side	13.0	33.0	cm^3/min
Unaccounted	3.6	9.2	cm^3/min

Discussion and Conlusions

Theoretical performance

As discussed before the specific resistance of the membrane should be approximately 150mΩ.cm^2, with the electrode polarizations in the same order of magnitude. Assume therefore for the total resistance at 800°C a value of 300mΩ cm^2, then at a Nernst voltage of 200 mV a current of i = 0.67 A/cm^2 would be expected. This translates into a flow by considering the reaction equations, Eqs.3 and realizing that 1 A = 1 C/s. Since each H_2 yields 2 electrons then with Faraday's constant (F = 96485 C/mol), the molar flow of H_2 follows from:

$$\phi_{H2} = \frac{i}{2F} = 3.45 \times 10^{-6} \text{mol}/(\text{s cm}^2) = 4.9\text{cm}^3/(\text{cm}^2 \text{min}) \qquad \text{Eq.7}$$

At 800°C no change in the CO/CO2 composition above the accuracy of the GC(±0.2%) could be measured. At 860°C an area specific flow of 0.4 cm^3/(cm^2 min) was measured which is about an order of magnitude lower than expected. At 1000°C the yield comes close to the number of Eq.7, however at this temperature polarization and resistivity should be significantly lower and the theoretical yield should be 4 to 5 times higher.

Electrode problem

As already indicated, the active layer Ni-CGO layer on the H_2/steam side seems to have disappeared. An EDAX analysis showed no Ce near the membrane. At high magnification it was found that the fine Ni from the active layer had migrated approximately 50-100μm into the substrate which probably diluted the Ni-CGO to such an extent that the Ce did not show on the EDAX anymore. It was also suggested that reactions had taken place, though a simple test where a new bisque fired tube was dipped in electrode material and then dried revealed after cross sectioning the same problem.

Interpretation of test data

The hydrogen cross-over to the CO/CO_2 side varied from 0.1% - 3% according to the GC analysis. At lower temperature (800°C), cross over was 0.1 - 0.2%, but as the temperature was raised the cross over increased to 3% at 1000°C. Post-test it was established that the glass – cement seal was leaking and that the membrane was intact. It was also established that the composition changed little if the H2 flow was reduced by 50%. It is therefore reasonable to assume that the observed hydrogen is mostly coming from the water that is crossing with the hydrogen as a result of the shift reaction. The high CO concentration would certainly favor this to happen. So to establish what the CO concentration before the shift was the H_2 concentration has to be added to it (Eq.4). Conversely to establish the CO_2 concentration before shifting the H_2 concentration has to be subtracted. Thus a CO/CO_2 ratio is found that is higher than when the correction is not applied and results in a lower yield number for the hydrogen evolution.

A similar analysis can be performed on the H_2/steam exit which to some extent, explains the discrepancies noted in Table 2.

Conclusions

- It was shown that chemically driven dissociation of water can be accomplished using two phase MIEC, consisting of a mixture of YST and 8YSZ;
- Hydrogen evolution rates of 0.4-1.8 cm^3/cm^2/min were measured in the temperature range of 860-1000°C;
- The membrane was dense post-test and appeared to have sufficient volume fractions of both conducting phases;
- The ionic and electronic phases do not appear to react, indicating a stable structure.

References

1. F. C. Gotzmann, R. Prasad, US patent 5733435 (1998).
2. T. J. Mazanec, T. L. Cable, J. G. Frye, Solid State Ionics, 53 – 56, (1992), p111.
3. K.W. Browall and R.H. Doremus, J. Am. Ceram. Soc. 60 (1977) p262.
4. H. Naito, H. Arashi, Solid State Ionics 79 (1995) p366.
5. U. Balachandran, T. H. Lee, S. Wang and S. E. Dorris, Int. J. Hydrogen Energy 29 (2004) p291.
6. S. Gopalan, JOM, 54, 5 (2002) p26.
7. U. Balachandran; E.S. Dorris, A.C. Bose, G.J.. Stiegel, T. H. Lee, US Patent 6,468,499.
8. J. Yoo, A. Verma and A. Jacobson, in: Ionic and Mixed Conducting Ceramics IV, ed. T Ramanarayanan, The Electrochemical Society, 28, (2001), p27.
9. S. Hui and A. Petric, J..Electrochem. Soc., 149, 1, (2002), J1-10.
10. O. A. Marina, N. L. Canfield, J. W. Stevenson, Solid State Ionics, 149, 2002, 21-28.
11. G. Pudmich, W. Jungen, F. Tietz, in: SOFC VI, ed. S. C. Singhal, M. Dokiya, The Electrochemical Society, 99-19, (1999), p577.
12. H. He, Y. Huang, J. M. Vohs, R. J. Gorte, Solid State Ionics, 175 (2004), 171-176.
13. O. A. Marina, L. R. Pederson, in:5th European SOFC Forum (1-5 July 2002), p481.
14. O. A. Marina, SECA Core Technology Program Peer Review, Boston, MA, May 11-13, 2004.
15. Y. Yoo, S. Koutcheko, A. Petric, in:5th European SOFC Forum (1-5 July 2002), p507.
16. S.. Koutcheko , Y. Yoo, A. Petric in:5th European SOFC Forum (1-5 July 2002), p655.

On a Explicit Relation for Plasma Layer Thickness

Dr. Kal Renganathan Sharma PE
Formerly Principal, Sakthi Engineering College
Anna University Affiliated, Chennai, Tamil Nadu, India
Phone: 91 044 2521849
Email: jyoti_kalika@yahoo.com

Abstract

The marginal zone theory is used to account for the observed Fahreus Lindquist effect when the viscosity of blood changes with the diameter of the capillary. An attributable cause is the axial accumulation of cells. The discharge rate from Hagen Poiseulle law at steady state was derived by Haynes (1960) for the core and plasma layer and a total discharge rate was expressed as a function of the pressure drop along the capillary, quartic dependence on the radius of the capillary and quartic dependence on the dimensionless marginal zone thickness. The apparent of viscosity of the blood is expressed as a function of the ratio of the core layer viscosity and the plasma layer viscosity. In order to back out a marginal zone thickness from a given set of information, the Charm and Kurland expression (1974) for the viscosity and hematocrit variation and the temperature dependence parameter of the hematocrit alpha can be used to develop two transcendental equations and two unknowns. This is the recommended procedure used currently and the equations are quartic in sigma, the dimensionless ratio of marginal zone thickness with the radius of the capillary. In this study an alternate procedure is developed for solving for the marginal zone thickness explicitly. On examination of the temperature variation parameter, alpha, for a given temperature, the transcendental equation is linear zed by a linear regression between αH and H the hematocrit. The slope and intercept values are used in the expression of $\sigma = \sqrt{\alpha_c/\alpha_T}$ and the hematocrit expression $H_T = H_C \sigma^2$. The resulting quadratic in σ^2 can be solved for and the sigma values obtained explicitly.

Introduction

The Fahraeus- Lindquist effect [1] can be captured by the marginal zone theory proposed by Haynes (2). Tube flow of blood at high shear rates (> 100/sec) exhibits the dependence of diameter on the viscosity effect is the Fahreus –Lindquist effect. When the diameter of the tube decreases below 500 microns upto 4 – 6 microns, the viscosity of the blood also decreases. The marginal zone theory may be used to characterize the effect from 4-6 microns to 500 microns in tune diameter. An expression is obtained for the apparent viscosity in terms of the plasma layer thickness, tube diameter and the hematocrit.

The bold flow within a tube or vessel is divided into two regions; a central core that contains the cells with a viscosity, μ_c, and the cell free marginal or plasma layer that consists only of plasma with a thickness of δ, and a viscosity equal to that of the plasma denoted by μ_p. In each region the flow is considered to be Newtonian and at steady state. For the core region, the governing equation may be written as;

$$\tau_{rz} = (\Delta P)r/2L = -\mu_c \partial v_z^c/\partial r \tag{1}$$

The boundary conditions can be written as;

$$\text{BC1: } r = R - \delta, \quad \tau_{rz}\ (\text{core}) = \tau_{rz}\ (\text{plasma}) \tag{2}$$

$$\text{BC 2: } r = 0, \quad \partial v_z^c/\partial r = 0 \tag{3}$$

The first boundary conditions stems from the continuity of the transfer of momentum across the interface between the core and plasma layer and the second boundary condition derives from the fact that the axial velocity would be a maximum at the center of the tube from symmetry arguments. In a similar vein, for the plasma layer the governing equation and boundary conditions can be written as;

$$\tau_{rz} = (\Delta P)r/2L = -\mu_p \partial v_z^p/\partial r \tag{4}$$

$$\text{BC 3: } r = R, \quad v_z^p = 0 \tag{5}$$

$$\text{BC 4: } r = R - \delta, \quad v_z^p = v_z^c \tag{6}$$

The boundary condition 3 comes from the zero velocity condition at the wall and the 4th boundary condition is that the velocity need be the same at the interface of the two phases. Eq. [1] and [4] may be integrated and the core and plasma flow rates given by the following;

$$Q_p = \pi(\Delta P)/8\mu_p L \,(R^2 - (R - \delta)^2)^2 \tag{7}$$

$$Q_c = \pi(\Delta P)\, R^2/8\mu_p L \,((R - \delta)^2 - (1 - /\sigma^2) (R - \delta)^4/ R^2 - \mu_p/\mu_c (R - \delta)^4/ 2R^2) \tag{8}$$

The total discharge rate of the blood would equal the sum of the flow rates in the core and plasma regions and is given by;

$$Q = \pi(\Delta P)\, R^4/8\mu_p L \,(1 - (1 - \delta/R)^4 (1 - \mu_p/\mu_c)) \tag{9}$$

Eq. [8] can be used to fit the apparent viscosity data and obtain values of the plasma layer thickness and the core hematocrit as a function of the tube diameter. A relation between the core hematocrit, H_c and the feed hematocrit, H_F and the thickness of the plasma layer is needed. An equation is needed to describe the dependence of the blood viscosity on the hematocrit since the value of the H_c will be larger than H_F because of the axial accumulation on the RBCs. This relative increase in the core hematocrit will make the equation in the core have a higher viscosity than the blood in the feed. The following equation by Charm and Kurland [3] may be used to express the dependence of the viscosity of blood at high shear rates on the hematocrit and temperature;

$$\mu = \mu_p (1/(1- \alpha H)) \tag{10}$$

$$\text{or } \alpha H = 1 - \mu_p/\mu \tag{11}$$

where

$$\alpha = 0.070 \exp(2.49\, H + 1107/T \exp(-1.69H)) \tag{12}$$

where the temperature is in Kelvin. Theses equations are valid to a hematocrit of 0.6 with a stated accuracy of 10 %. If a subscript occurs on the viscosity then the corresponding values of H and α in the above equations will carry the same subscript. If $\sigma = 1 - \delta/R$ then the solution for the plasma layer thickness are implicit and require solution of two equations and two unknowns:

$$\mu_{app}/ \mu_F = (1 - \alpha_F H_F) /(1 - \sigma^4 \alpha_C H_C) \tag{13}$$

$$H_C/H_F = 1 + (1 - \sigma^2)^2/\sigma^2 (2 (1 - \sigma^2) + \sigma^2 \mu_p/\mu_c) \tag{14}$$

An explicit expression for the plasma layer thickness is desirable and a method is explored in this study.

Methodology

Eq. [12] is examined using a spreadsheet. It can be observed that αH is linear with H at a given temperature for the range of hematocrit for which the Charm and Kurland expression is valid (Figure 1). Thus for the core and tube ;

$$\alpha_c H_c = m H_c + C \qquad (15)$$

$$\alpha_T H_T = m H_T + C \qquad (16)$$

The slope and intercept can be obtained by a least squares regression between the αH and H as given by Eq. [12]. From a balance of cells in the two phases in the tube it can be seen that;

$$H_T = \sigma^2 H_c \qquad (17)$$

Dividing Eq. [14] by [15],

$$\alpha_c H_c \,/\, \alpha_T H_T = m H_c + C / m H_T + C \qquad (18)$$

From Eq. [11] it can be seen that;

$$\alpha_T H_T = 1 - \mu_p / \mu_{app} \qquad (19)$$

From Eq. [9] with minor rearrangement it can be seen that;

$$\alpha_C H_C \sigma^4 = 1 - \mu_p / \mu_{app} \qquad (20)$$

Equating [19] and [20] and combining with Eq. [17],

$$\sigma = \sqrt{\alpha_T / \alpha_C} \qquad (21)$$

Plugging Eq. [21] and [17] into Eq. [18];

$$\sigma^4 = (m H_T + C) / (m H_T / \sigma^2 + C) \qquad (22)$$

with, $\sigma^2 = p$

the quadratic can be solved;

$$C p^2 + (m H_T - 1) p + C = 0$$

Thus an explicit expression for the plasma layer thickness in terms of the tube hematocrit is developed. The tube hematocrit can be read from the linear regression developed between the αH and H at a given temperature once the apparent viscosity of the tube is known.

References

[1] R. L. Fournier, 1999, *Basic Transport Phenomena in Biomedical Engineering*, Taylor & Francis, Philadelphia, PA, USA.

[2] R. F. Haynes, 1960, Physical Basis of the Dependence of Blood Viscosity on Tube Radius, Am. J. Physiol., 198, 1193.

[3] S. E. Charm, G. S. Kurland, 1974, *Blood Flow and Microcirculation*, John Wiley, New York, NY, USA.

Figure 1: Linearization of the Charm and Kurland Expression

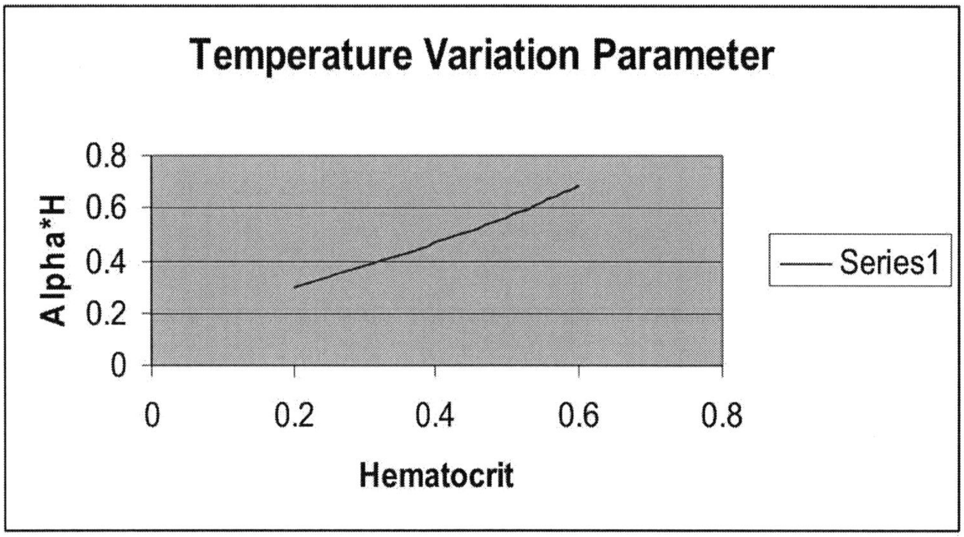

121b The Influence of Bend Geometry on the Performance of a MicroGC

Adarsh D Radadia, Keith R Cadwallader, Mark A Shannon, Rich Masel

A definite need for fast, reliable, handheld, low power equipments for environmental monitoring is realized in many areas. A Micro-GC with on-chip devices is a strong potential solution. The separation column forms an important part of the GC where segregation of analytes occur. For quick and efficient separation of analytes it is very important to know how micro-columns need to be designed. Bends are unavoidable features in microfluidic systems due to design or technological constraints. However, these bends induce pressure drop and dispersion that decreases the separation time and efficiency. This paper shows how microfabricated columns with different bend designs perform differently. The following designs were fabricated on chip with DRIE technology: simple circular bends, faceted designs with expansion at bends, faceted designs with contraction at bends, a contraction-expansion design and a contraction-expansion design with race-track compensation. Different designs give different degrees of separation. A pressure drop comparison is made using the obtained carrier mass flow rate at constant inlet pressure. Dispersion is characterized using the peak asymmetry from an FID detector signal. The results indicate that dispersion in the contraction-expansion design with racetrack compensation is minimum at the expense of pressure drop.

EFFECTS OF INLET GEOMETRY ON GAS-LIQUID TWO-PHASE FLOW IN MICROCHANNELS

Masahiro Kawaji[1], Koji Mori[2], Daniel Bolintineanu[1] and Y. Shibata[3]

1. Dept. of Chemical Engineering and Applied Chemistry
University of Toronto
200 College Street
Toronto, Ontario M5S 3E5, Canada
E-mail: kawaji@ecf.utoronto.ca
2. Dept. of Mechanical Engineering
Osaka Electro-Communications University
Osaka, Japan
3. Dept. of Mechanical Engineering
Ibaraki National College of Technology,
Hitachinaka, Ibaraki, Japan

INTRODUCTION

Although few two-phase flow studies are available in the literature that involve a channel diameter of 100 μm or less, past results indicated wide differences in the adiabatic two-phase flow characteristics such as flow patterns, void fraction and pressure drop data, possibly due to the inlet configuration.

Air-water two-phase flow experiments were conducted in 25 μm and 100 μm circular microchannels by Feng and Serizawa (1999), Serizawa and Feng (2001) and Serizawa et al. (2002). Five flow patterns were observed: dispersed bubbly flow, gas slug flow, liquid-ring flow, liquid lump flow, and liquid droplet flow. The occurrence of liquid-ring flow and liquid lump flow clearly indicated that there are differences in the flow patterns between microchannels of $D \leq$ 100 μm and minichannels of $D \geq 1$ mm.

Kawahara et al. (2002) and Chung and Kawaji (2004) investigated the two-phase flow of nitrogen gas and water through horizontal circular microchannels of 50 ~ 250 μm diameter. They reported significant differences in the flow regime maps and void fraction data from those previously described for minichannels with diameters above ~ 1 mm. In particular, they reported the existence of flow patterns unique to microchannels (liquid-ring flow and serpentine-like gas core flow) and deviation of void fraction data from an Armand-type correlation normally applicable to minichannels with diameters ~ 1 mm. This was in contrast with the void fraction data of Serizawa et al. (2002) for a 20 μm circular microchannel, which could be described by an Armand-type correlation. Chung et al. (2004) also found that the two-phase flow characteristics in a 96 μm square channel and a 100 μm circular channel were quite similar and independent of the channel cross section for a mixture of nitrogen gas and water. Furthermore, Kawahara et al. (2005) found only a moderate effect of fluid properties such as liquid viscosity and surface tension on the two-phase flow characteristics in a microchannel.

Thus, the results in the literature are not entirely consistent and the effect of reducing the channel diameter on two-phase flow in microcahnnels remains uncertain. Some results for microchannels differed from those in conventional channels, while others stayed the same. Bubbly and droplet flow may not always occur in microchannels. The void fraction data in microchannels may deviate from the Armand (1946) correlation and the two-phase friction pressure drop in microchannels could be predicted by different models.

Hence, the objective of this work was to identify the key reason for different two-phase flow characteristics for the same microchannel of ~ 100 μm diameter. In this paper, the effects of inlet configuration and the method of gas-liquid injection are explored and explained.

EXPERIMENTAL APPARATUS AND DATA ANALYSIS

The experimental setup and procedure are explained below briefly, and the reader can refer to Kawahara et al. (2002) and Chung et al. (2004) for details. Figure 1 illustrates the experimental apparatus used. De-ionized water was driven through the horizontal microchannel by a pneumatic-type pump after being mixed with a nitrogen gas. A gas-liquid mixture was pumped into a given microchannel positioned horizontally.

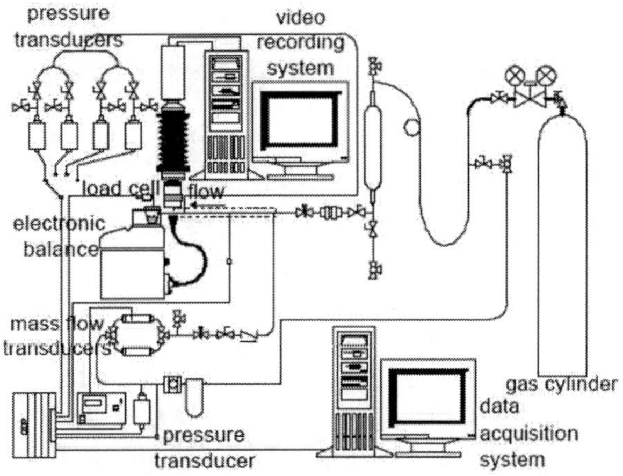

Fig. 1 Experimental apparatus

Figure 2 illustrates the gas-liquid inlet section used in the present work. A micro 4-way junction with an inner bore diameter of 100 μm was used to directly connect the 100 μm diameter microchannel test section to the gas and liquid injection lines. The fourth line was connected to a pressure transducer to measure the test section inlet pressure. In the present work, liquid was injected into the main channel and gas injected into the branch, and mixing of the gas and liquid phases took place inside the 100 μm diameter 4-way junction immediately before the test section. Thus, there was no change in the tubing diameter between the gas-liquid mixing section and the test section. A circular microchannel (capillary tubing with standard polyimide coating from Polymicro Technologies) with an inner diameter of 100 μm and 64 mm length was used for the test section. A high *L/D* ratio (*L/D* = 640) was involved to diminish the entrance and exit effects. The window for flow visualization was located as far back as possible on the microchannel to achieve fully-developed flow.

In the previous experiments, gas and liquid phases were pre-mixed in the upstream mixing section which consisted of an internal core passage for liquid and an external annular passage for gas. Liquid flowed axially through a 500 μm tube, while gas flowed peripherally through a 200 μm gap. Eventually, the two streams mixed by flowing collectively into a 500 μm ID/1/16" OD tube that was connected to the 250 μm ID tee-junction for pressure measurements upstream of the test section. Beyond the tee fitting, the newly formed two-phase flow entered the microchannel. The two-phase mixture thus flowed through 750 and 500 μm diameter tube sections before entering the 250 μm diameter T-junction and finally into the 100 μm diameter test section.

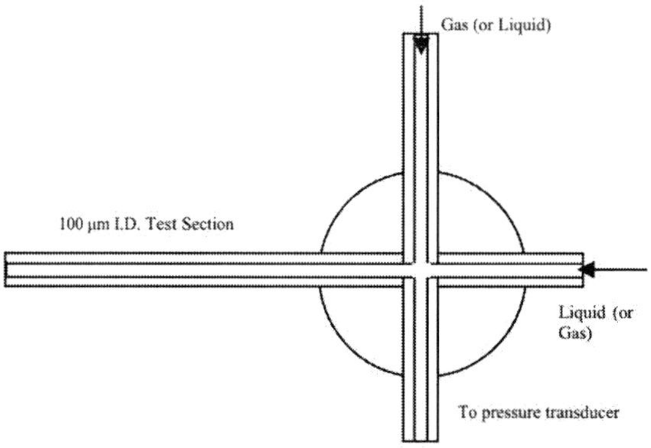

Fig. 2 Gas-Liquid mixing upstream of the test section in new experiments

Three pressure transducers covering different ranges (3450 ± 8.6 kPa, 210 ± 0.52 kPa and 24.9 ± 0.037 kPa,) were used to measure the pressure drop between the microchannel inlet and outlet. Three additional pressure transducers of various ranges (13790 ± 34 kPa, 170 ± 2 kPa and 1380 ± 3.4 kPa) were used for single-phase liquid flow. The liquid was discharged freely from the test section, so that the pressure was atmospheric and temperature ambient at the microchannel outlet. The liquid flow rate was determined by using an electronic balance (120 ± 0.001 g) or load cell (50.97 ± 0.255 g) to weigh the discharge liquid collected over time. The gas flow rate was measured using three mass flow meters to encompass a wide range of conditions (1 ± 0.01 sccm, 10 ± 0.1 sccm and 50 ± 0.5 sccm). The analog signals for all of the pressure, temperature and mass flow rate readings were sampled at 100 Hz by a 16-bit data acquisition system.

Background illumination to see the flow patterns was generated by a cold lamp located behind the test section. The view field was magnified by attaching a 5X microscope objective lens to the video camera. Two monochrome CCD cameras were used interchangeably. The first video camera had a resolution of 1024(H) x 1024(V) pixels and recorded at a maximum frame rate of 30 fps and shutter speed of 1/16,000 s. The second video camera had a resolution of 648(H) x 484(V) pixels and recorded at a maximum frame rate of 125 fps and minimum exposure period of 66 μs.

The experiments covered the following flow rate ranges: $0.027 \leq j_L \leq 2.05$ m/s for the superficial liquid velocity and $0.1 \leq j_{G,AVG} \leq 50$ m/s for the average superficial gas velocity. Initially, experiments were conducted with de-ionized water to calculate the single-phase friction factor from the frictional pressure drop. Afterward, the computed friction factor was compared with the conventional value for laminar flow in a pipe. The two-phase frictional pressure drop was calculated by subtracting estimates of the pressure drop due to inlet contraction and flow acceleration from the total pressure drop measured in the horizontal channel. To estimate the pressure drop attributed to acceleration of the flow, the experimental void fraction evaluated from the analysis of the video recordings was used.

RESULTS AND DISCUSSION

The experimental results on the two-phase flow patterns, void fraction and two-phase frictional pressure drop obtained are presented below and compared with those reported previously by Kawahara et al. (2002) to determine the effect of inlet geometry and the gas-liquid mixing method.

Two-Phase Flow Patterns

The two-phase flow patterns obtained in the new experiments are found to be quite different from those observed in our previous work. Typical examples of the flow patterns seen in the new experiments are shown in Fig. 3 below.

(a) Plug flow

(b) Plug flow with two gas plugs separated by a short liquid slug

(c) Slug flow with a ring-shaped liquid film around a gas core

Fig. 3 Two-phase flow patterns observed in the new experiments

In the previous experiments using a series of reducer sections between the gas-liquid mixer and test section, short gas bubbles or plugs like those shown in Fig. 3 were seldom observed. Instead, slug flows with a long gas core surrounded by a thin, thick or wavy liquid film were predominant. Thus, the flow patterns observed in the new experiments were classified differently from the previous experiments and according to the following definitions.

Bubbly flow - the presence of spherical bubbles whose diameters are less than the diameter of the microchannel

Plug flow - the presence of gas plugs that are shorter than the width of the viewing window (~250 μm), and are separated by liquid bridges

Slug flow - the presence of gas slugs that are longer than the viewing window (i.e. no image captures a full gas slug)

Semi-annular flow - gas core flow surrounded by a liquid film and liquid slugs are seen infrequently

Ring flow - 10% or more of the images for a given run have a ring-shaped liquid film surrounding the gas core

The two-phase flow pattern maps were then constructed based on the above definitions for the two gas-liquid injection schemes as shown in Fig. 4. When the inlet section consisted of a series of reducer sections from 750 to 100 μm, the flow patterns obtained in the previous experiments by Kawahara et al. (2002) were predominantly those of slug flows with long gas cores surrounded by thin, thick or ring-shaped films. Depending on the frequency of the appearance of different liquid film shapes, the flow patterns were defined as slug-ring flow, ring-slug flow and multiple flow, and a two-phase flow pattern map was constructed as shown in Fig. 5. Although a semi-annular flow pattern was observed in both previous and new experiments, a strong difference exists between the two maps in the appearance of short gas bubbles and plugs. This is due to the fact that small gas bubbles are readily created at the T-junction. The mechanism of small gas bubble generation at the T-junction can be readily imagined as shown in Fig. 6. Such a flow mechanism was investigated recently by Shibata et al. (2003). Gas bubbles are generated as a result of interfacial instability downstream of the T-junction as shown in Fig. 7.

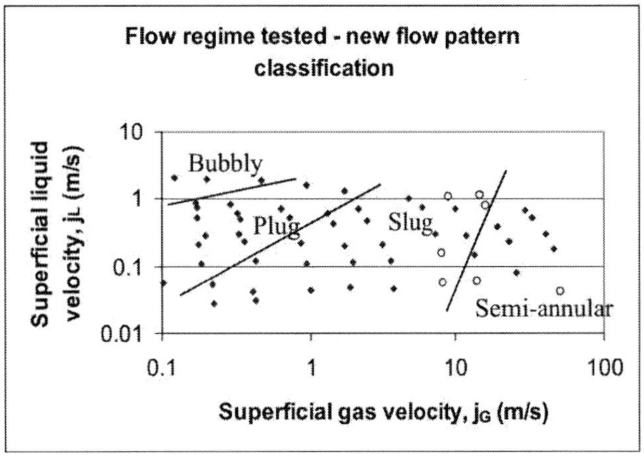

Fig. 4 **Two-phase flow pattern maps for new experiments** (o indicates the presence of a ring-film flow pattern)

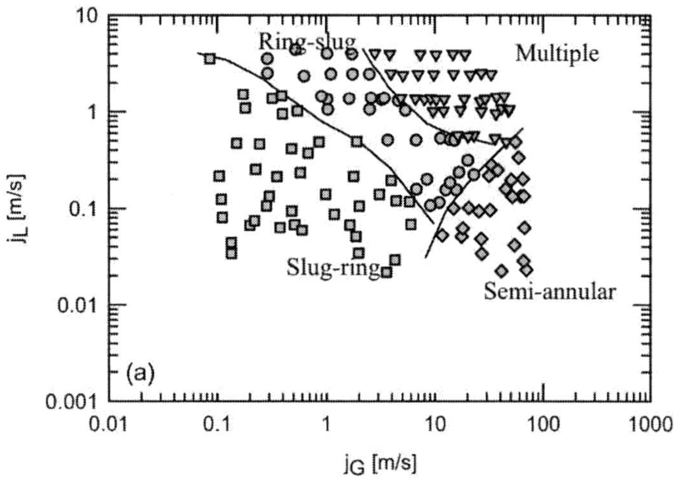

Fig. 5 Two-phase flow pattern map obtained in previous experiments by Kawahara et al. (2002)

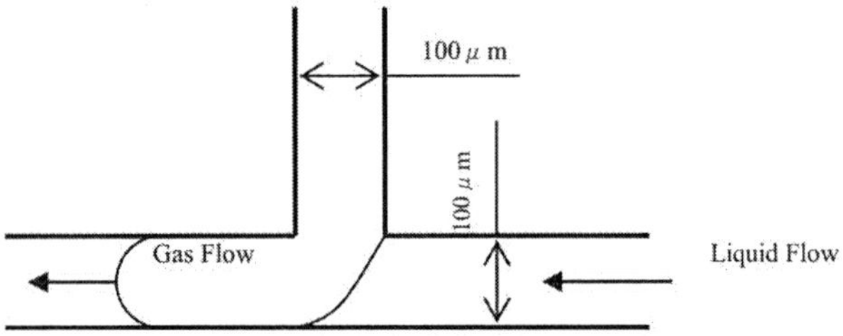

Fig. 6 Gas bubble generation mechanism at the T-junction

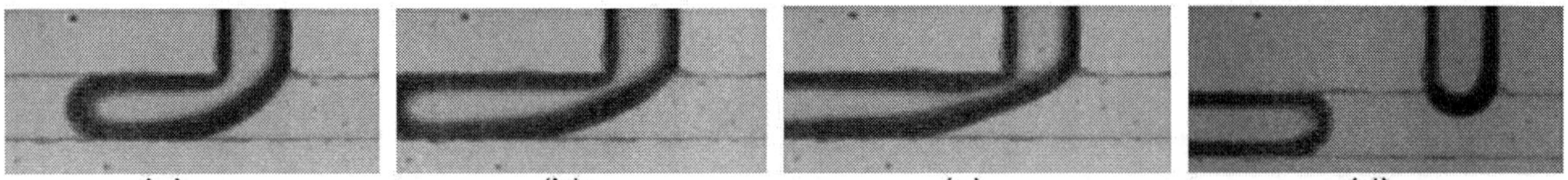

(a) (b) (c) (d)

Fig. 7 Images showing gas plug generation at a T-junction (Shibata et al., 2003)

The use of a micro T-junction to inject the gas and liquid into a microchannel resulted in two-phase flow patterns in the microchannel downstream that are very different from those observed by Kawahara et al. (2002) when the gas and liquid were injected into a much larger inlet section first and then flowed into the microchannel through a series of reducer sections. The larger inlet cross section (D = 750 μm) would allow larger gas plugs to form upstream of the microchannel test section. When these gas plugs flow into the microchannel with a much smaller diameter (D = 100 μm), the gas slug length would increase by more than 50 times. So, for example, if the length of the gas plug formed in the inlet section were 2 mm, then it would form a 100 mm long gas slug in the microchannel. This can explain why Kawahara et al. (2002) obtained mostly long gas core flows rather than short gas plug flows.

The differences in the flow patterns obtained at the same gas and liquid superficial velocities can cause a large difference in the void fraction as discussed below.

Void Fraction

To estimate the void fraction in the 100 μm microchannel, image analysis was performed on about 200 video images in each run. The instantaneous void fraction was estimated for plug flow images such as those shown in by measuring the length of a gas bubble or plug, and calculating the volume occupied by assuming a body of revolution. The images were assigned a void fraction of zero when they depicted liquid flowing alone in the entire view window or a void fraction of unity when they depicted gas core flow with a thin-smooth liquid film or ring-shaped liquid film. For a thick-smooth liquid film, the volume fraction of gas was estimated by squaring the ratio of the gas core radius to the channel radius (Kawahara et al., 2002). The effect of optical distortion on the above-mentioned method of measuring the void fraction, particularly for gas core flows with ring-shaped films and thick liquid films, was already addressed by Kawahara et al. (2002).

Figure 8 presents the time-averaged void fraction, ε, data plotted against the volumetric quality, β ($= \frac{J_G}{J_G + J_L}$), for the new and previous experiments, respectively. The homogeneous flow model ($\varepsilon = \beta$) and the Armand correlation (1986) for narrow channels recommended by Ali et al. (1993) are also shown in the figures, respectively. Ali et al. (1993) recommended using a correlation of the form ($\varepsilon = 0.8\beta$) originally developed by Armand (1946) for narrow rectangular channels with D_H ~1 mm.

The new void fraction data show a linear relationship between the void fraction and volumetric quality, in agreement with the homogeneous flow model predictions. On the contrary, the void fraction data from the previous experiments by Kawahara et al. (2002) show a non-linear void fraction-to-volumetric quality relationship given by,

$$\varepsilon = \frac{C_1 \beta^{0.5}}{1 - C_2 \beta^{0.5}} \qquad [1]$$

where C_1 = 0.03 and C_2 = 0.97 for the 100 μm circular channel. Equation (1) indicates void fraction initially increasing slowly with increasing volumetric quality and then exponentially after

β =~0.8. This trend implies a strong departure from homogeneous flow and indicates a large velocity slip between the liquid and gas phases.

It is obvious that as the inlet geometry is changed, the void fraction is influenced strongly because of the changes in two-phase flow patterns. A plug flow pattern with alternating flows of short gas plugs and liquid slugs seen in the new experiments would lead to slip ratios getting close to unity representative of homogeneous flow, while a fast gas core flow with slow liquid film flow would result in high slip ratios and low void fractions.

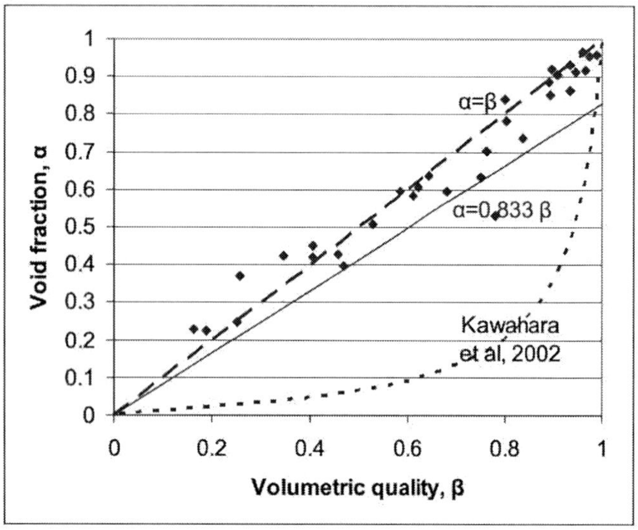

Fig. 8 Void fraction data obtained in the new experiments

Friction Pressure Drop

The two-phase friction pressure drop was evaluated from the measured total pressure drop by subtracting the acceleration pressure drop. The data are correlated using the Lockhart-Martinelli correlation (1949), which was developed based on a separated flow assumption. This correlation uses a two-phase friction multiplier, ϕ_L^2 and a Lockhart-Martinelli parameter, X,

$$\phi_L^2 = 1 + \frac{C}{X} + \frac{1}{X^2} \qquad [2]$$

where the coefficient, C, is a constant ranging in value from 5 to 20 in larger channels, depending on whether the liquid and gas flows are laminar or turbulent. The C-value in Eq. [2] is known to decrease as the channel hydraulic diameter is reduced (Kawaji, 1999).

Comparisons of the new and previous two-phase friction multiplier data with the predictions of the Lockhart-Martinelli correlation are shown in Fig. 9. Curves given by Eq. 2 with C = 5 and best fit values of C = 2.3 and 2.5 for the new experiments and C = 0.24 for the previous experiments (Kawahara et al., 2002) are also shown. As in the void fraction data, the friction pressure drop data show significant differences between the new and previous experiments indicating the effect of the inlet geometry.

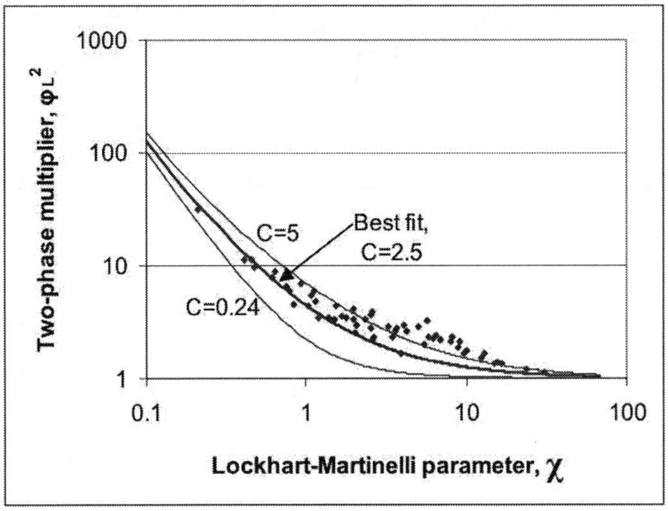

Fig. 9 Two-phase friction multiplier data for the new experiments

CONCLUSION

The effects of inlet geometry and gas-liquid injection method on adiabatic two-phase flow in a 100 μm microchannel have been investigated experimentally. Nitrogen gas and water were injected into a micro T-junction with the same diameter as the microchannel connected downstream. The two-phase flow patterns observed have been found to differ strongly from those obtained in a previous study using a 750 μm diameter mixing section. With the micro T-junction at the inlet, plug and slug flows with short gas plugs and liquid slugs were observed. Due to the plug and slug flow patterns, the void fraction data for the microchannel agreed with the homogeneous flow model, which was entirely different from the data obtained with a 750 μm diameter mixing section. Two-phase friction pressure multiplier data could be well correlated in terms of the Lockhart-Martinelli parameter but the value of the constant in Chisholm's correlation was much larger than the value obtained previously.

ACKNOWLEDGMENTS

This work was supported by a Discovery research grant from the Natural Sciences and Engineering Research Council of Canada. K. Mori would like to thank Osaka Electro-Communications University for a research leave grant, which allowed a one-year stay at the University of Toronto.

REFERENCES

Ali, M.I., Sadatomi, M., and Kawaji, M., 1993. Two-phase flow in narrow channels between two flat plates. *Can. J. Chem. Eng.*, 71(5), 657-666.

Armand, A.A., 1946. The resistance during the movement of a two-phase system in horizontal pipes, Izv. Vses. Teplotekh. Inst., 1, 16-23 (AERE-Lib/Trans 828).

Chung, P.M.Y., and Kawaji, M., 2004. The effect of channel diameter on adiabatic two-phase flow characteristics in microchannels. *Int. J. Multiphase Flow*, Vol. 30, No. 7-8, pp. 735-761.

Chung, P.M.-Y., Kawaji, M., Kawahara, A., and Shibata, Y., 2004. Two-phase flow through square and circular microchannels – effect of channel geometry, *ASME J. Fluids Engineering*, Vol. 126, No. 4, pp. 546-552.

Feng, Z.P., and Serizawa, A., 1999. Visualization of two-phase flow patterns in an ultra-small tube, Proc. of the 18th Multiphase Flow Symposium of Japan, July 15-16, Suita, Osaka, Japan, 33-36.

Kawahara, A., Chung, P.M.-Y., and Kawaji, M., 2002. Investigation of two-phase flow pattern, void fraction and pressure drop in a microchannel. *Int. J. Multiphase Flow*, 28(9), pp. 1411-1435.

Kawahara, A., Kawaji, M., Chung, P.M.-Y., Sadatomi, M. and Okayama, K., 2005. Effects of channel diameter and liquid property on void fraction in adiabatic two-phase flow through microchannels. *Heat Transfer Engineering*, in press.

Kawaji, M., 1999. Fluid mechanics aspects of two-phase flow, in Handbook of Phase Change: Boiling and Condensation, Chapter 9, Taylor & Francis, pp. 205-259.

Kawaji, M. and Chung, P.M.-Y., 2004. Unique characteristics of adiabatic gas-liquid flows in microchannels: Diameter and shape effects on flow patterns, void fraction and pressure drop, *Journal of Microscale Thermophysical Engineering*, Vol. 8, No. 3, pp. 239-257.

Lockhart, R.W., and Martinelli, R.C., 1949. Proposed correlation of data for isothermal two-phase, two-component flow in pipes. *Chem. Eng. Progr.*, 45(1), pp. 39-48.

Serizawa, A., and Feng, Z.P., 2001. Two-phase flow in micro-channels. Proceedings of the 4th International Conference on Multiphase Flow, May 27-June 1, 2001, New Orleans, LA, USA.

Serizawa, A., Feng, Z., and Kawara, Z., 2002 Two-phase flow in microchannels. *Exp. Therm. Fluid Sci.*, 26(6-7), 703-714.

Shibata, Y., Ikeda, K., Hanri, T., Chung, P.M.Y., and Kawaji, M., 2003. Study on Two-Phase Flow through a T-junction, Paper # FDSM2003-45385 in the Proc. of FEDSM'03, 4th ASME-JSME Joint Fluids Engineering Conference, July 6–11, 2003, Honolulu, Hawaii.

Pulsating Heat Pipes for Microelectronics Cooling

K. Nikkanen, C. Lu and M. Kawaji
Department of Chemical Engineering & Applied Chemistry
University of Toronto
200 College Street
Toronto, ON, M5S 3E5
Canada
E-mail: kawaji@ecf.utoronto.ca

INTRODUCTION

In 1965 Gordon Moore, co founder of Intel, remarked that the complexity of microchips had increased at a rate of a factor of 2 per year [1]. More importantly, he predicted that this trend would continue into the future. Complexity refers to the number of transistors on a chip; the increase of which has been largely helped by advances in the miniaturization of electronic elements. This prediction has been upheld, with slight modification of the time line, even now into the 21st century, reaching as many as 410 billion transistors packed onto a single chip [2]. The demand for even further miniaturization of electronic devices is leading to greater chip packing densities and thus higher chip power densities. Proper functioning of electronics requires strict temperature conditions [3]; therefore the new challenge in the electronics industry has become removing large quantities of heat quickly and efficiently from a chip in order to keep chip temperatures within an operational range.

The use of heat sinks has been the traditional method of heat removal for electronic components. On-chip heat sinks enhance natural convection by conducting heat from a dimensionally small source to a large surface area. Having no moving parts, heat sinks require no maintenance and thus are more reliable over non-passive cooling strategies. The rising demand for heat removal due to miniaturization requires new solutions, however, as heat sinks alone are reaching their thermal limit [4]. Combination solutions which include heat sinks, fans, and heat pipes are becoming more common. The Pentium 4 cooling system for the Shuttle XPC PH4 [4] is such a hybrid cooling solution, and is comprised of a copper plate heat spreader bonded beneath the chip, a mounted heat sink, and 6mm diameter heat pipes. The heat pipes are sandwiched between the copper plate and heat sink, and conduct some heat away from the source to a stacked fin structure where it is removed by natural convection.

Heat pipes are efficient heat transfer devices which use the evaporation and condensation of a fluid to transport heat. The common heat pipe is a sealed, evacuated pipe containing a fluid. One end of the pipe is the evaporation region where heat is introduced to evaporate the fluid. The vapour then travels to the opposite end, the condensation region, where it loses its heat of vaporization and returns to liquid form. The walls of the heat pipe are often coated with a porous material that provides capillary pumping pressure to return the condensed fluid to the evaporation region leaving the core of the pipe as a vapour channel, thus opposing flows of vapour and fluid are permitted. An alternative to the tubular heat pipe is the pulsating heat pipe. A pulsating heat pipe consists of one continuous channel that meanders back and forth between the evaporation and condensation regions. The channel can be closed (looped) or open (un-looped). The channel diameter in a pulsating heat pipe is on the order of ~1mm hydraulic diameter so that stratified flow is prevented and a vapour plug/ liquid slug flow is formed.

There are several major differences between the tubular and pulsating heat pipes which are contrasted in figure 1. The primary difference is in the mechanisms of fluid and heat transport. In the pulsating heat pipe, only a small percentage of the total heat is used to evaporate the fluid. The remaining heat goes into raising the temperature of the liquid or vapour. The growth and contraction of vapour bubbles as they travel through the evaporation and condensation regions, respectively, creates the pumping power that pushes the fluid through the channel picking up and releasing the sensible heat [5]. The bubble pumping pressure and the one directional flow of the fluid eliminates the requirement of a wick structure making for a much simpler construction.

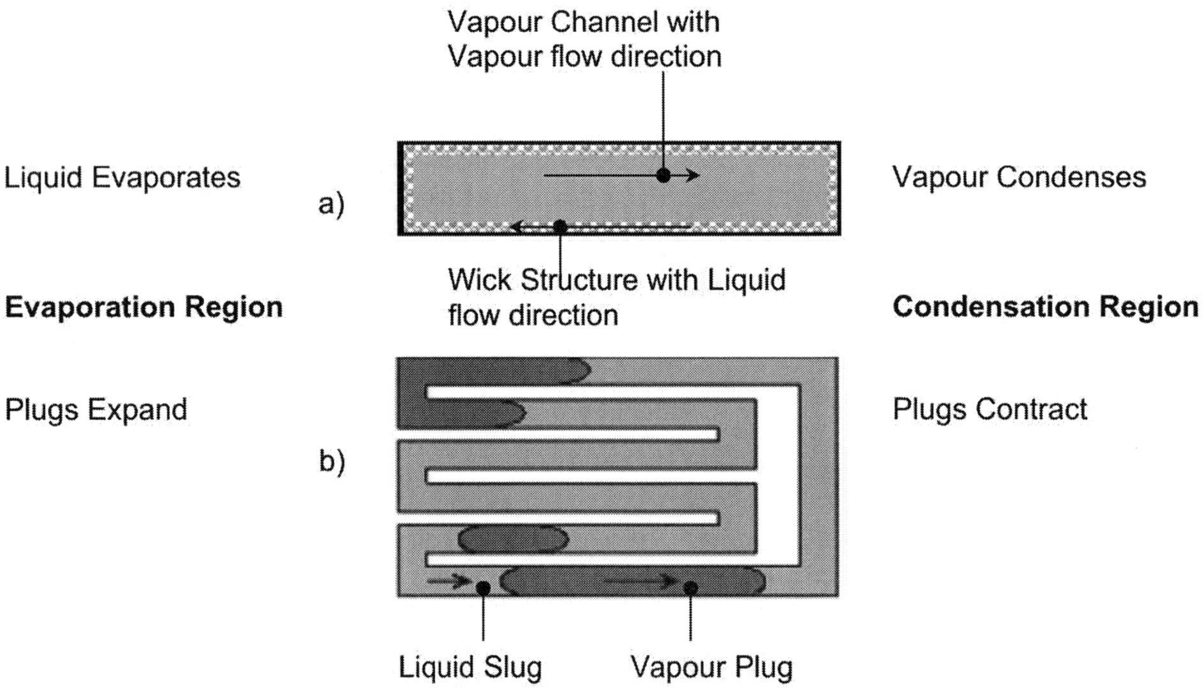

Figure 1 - Internal structures and flow mechanisms of the a) tubular heat pipe and b) looped pulsating heat pipe

It is believed that a pulsating heat pipe can have a heat transfer coefficient significantly higher than that of a tubular heat pipe due to the different heat transfer mechanism [5] and also because of the non opposing flow, the greater fluid volume, and the increased surface area for conduction of heat through the casing. Due to its benefits over tubular heat pipes, the pulsating heat pipes could serve as a very useful tool in the effort to remove large heat fluxes from modern electronic devices. The effect of gravity on the operation of a pulsating heat pipe has been studied previously by our group [6].

The objective of this work is to investigate the effects of heat pipe's structural parameters such as the shape and size of the channels and looped or unlooped configuration, as well as fluid medium, fill ratio and heat pipe orientation on the performance of a pulsating heat pipe.

EXPERIMENTAL

In order to design a pulsating heat pipe for optimal heat transfer it is important to have a better understanding of its operational characteristics. Physical features such as channel size and shape can have a significant effect on the propagation of vapour bubbles, and thus on the pumping power. To this end, experiments have been conducted on a variety of heat pipe designs in which structural parameters such as channel number, size, shape and configuration were varied. An overview of the different heat pipes tested is given in Table 1. The channel area is the cross sectional area, and the channel number means the number of channels in parallel that are connected in a serpentine configuration.

Table 1 Structural characteristics of test heat pipes

		Channel			Pipe	
Name	Looped	Shape	Area (mm^2)	Number	Width (mm)	Length (mm)
RL	Yes	Rectangular	2	11	20	135
RU	No	Rectangular	2	11	20	142
TL	Yes	Triangular	1	26	19	148
TU	No	Triangular	1	26	19	143

It was desirable to keep the construction of each heat pipe as simple as possible. For this reason aluminum multi-port extrusion tubing from Brazeway Inc. was chosen for the body of the heat pipe. The ends were machined and sealed to form the internal serpentine channel structure and an inlet was constructed from a septum-sealed tube fitting for easy evacuation and charging of the heat pipe via a fine gauge needle and a syringe. The rectangular channel looped heat pipe was constructed with a plastic end cap to accommodate the loop. This permitted the observation of vapour plug and liquid slug flows within the device. At high power settings the rapid travel of the vapour plugs could be easily seen and corresponded with successful operation.

The experimental set up consisted of an electric heater, interface material (thermal grease), the pulsating heat pipe, an extruded aluminum heat sink with fins, fan, thermocouples and data acquisition system. During operation the heater temperature and surface temperatures at various locations along the heat pipe were monitored. The physical arrangement including the general location of thermocouples and heat source is shown in figure 2. Successful operation of the heat pipe would result in a low temperature difference across the length of the heat pipe, and reduced steady state temperature of the heat source at each power setting when compared to operation of the heat source with a solid aluminum plate in place of the heat pipe.

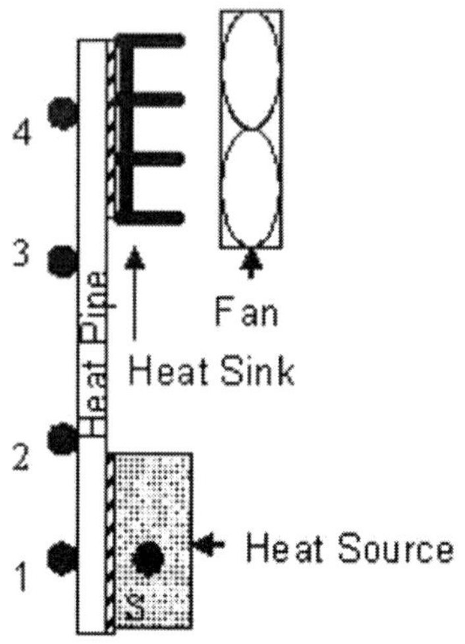

Figure 2 Side view of Experimental Setup (Thermocouple data are recorded by a data acquisition system and Labview program)

RESULTS

The effect of various operational parameters such as fill ratio, fluid medium and orientation were tested and compared for the different designs. This paper outlines some of the observations from the fill ratio and fluid medium comparison tests.

Changing the fill ratio of the heat pipe prompted a dramatic change in heat pipe performance. Figure 3 shows the effect of fill ratio on the performance of the rectangular channel looped heat pipe as measured by the maximum attainable power. The maximum attainable power is defined as the maximum power setting (in 5 W increments) at which the steady state heat source temperature remains below 100°C.

This particular heat pipe was shown to have an optimal fill ratio between 30 and 50%. A comparison of the heating profiles of this heat pipe, RL, at 50% and 10% fill ratios can identify the characteristics of a properly functioning heat pipe (figure 4).

The first sign of better efficiency is the lower heat source temperatures. Second, the temperature gradient along the heat pipe is almost negligible for a 50% fill ratio, but very significant for the 10% fill ratio tests. This gradient is seen as the temperature difference between thermocouple 1 which is located at the heat source, and thermocouple 3 positioned just below the heat sink.

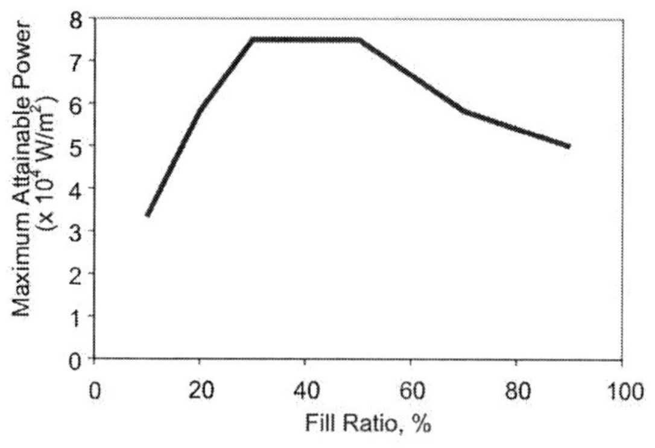

Figure 3 – Maximum Attainable Power for Rectangular Channel Looped Heat Pipe at Various Fill Ratios of De-ionized Water

Another important comparison is the transient behaviour of each test case. As power is increased in increments of 5W (once steady state has been achieved) the time that it takes to reach the new steady state temperature is significantly shorter in the 50% fill ratio case compared to the 10% case. This is important because electronic chips do not generate a steady heat flux and a cooling device must be able to respond to changes in heat flux in a timely manner.

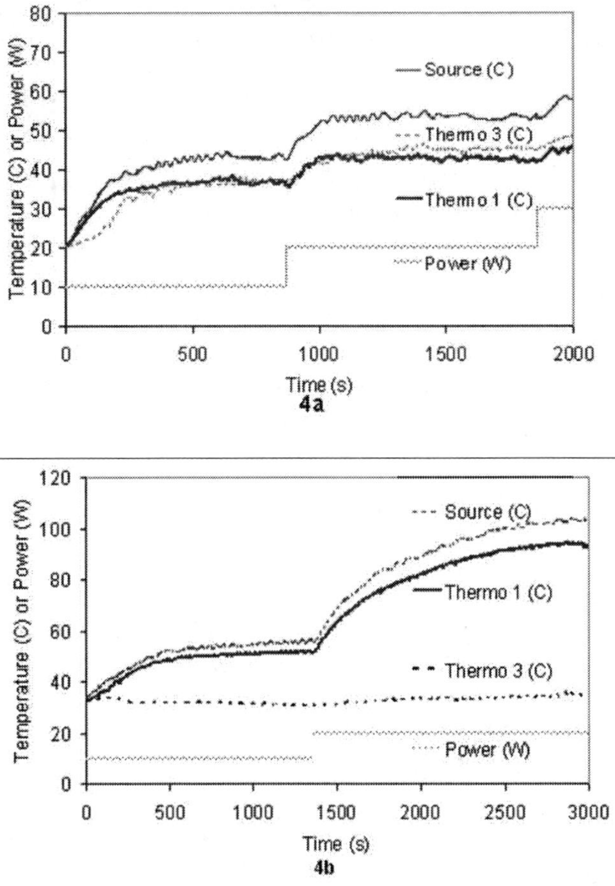

Figure 4 – Temperature and Power Profiles for RL at (a) 50% fill ration and (b) 10% fill ratio.

The source of the oscillation in heat source temperature values that is readily observed in the 50% fill ratio case and to a lesser extent in most other cases has not yet been identified. The oscillation has a period of approximately 30 seconds and amplitude of 1 ^{o}C. The corresponding disturbance in the temperatures at positions 1 and 3 in the 50% fill ratio test suggests that the heat pipe is able to respond well to the changing power. In the 10% fill ratio test, the temperature profiles for thermocouples 1 and 3 are fairly smooth in comparison, and thus not responding as quickly to the changing conditions in the heater power.

In comparison, the un-looped heat pipe RU using the same fluid was not able to function satisfactorily at all as can be seen in the fill ratio vs. maximum attainable power graph in figure 5a. Previous studies have indicated that un-looped heat pipes should work very well [7] which contradicts the present results for RU. Our hypothesis is that for an un-looped heat pipe to work a greater number of channels is required in order to reduce the importance of the edge channels in which vapour plug travel is restricted by a finite end. In effect, the central channels will act more like those in a looped configuration.

A triangular channel un-looped heat pipe, TU, was made to test this hypothesis. TU had more than twice the channels in the same amount of space as RU. As expected it performed better (figure 5b) than RU, though still not as well as the looped version, RL. To show that the increase in channel numbers helped to reduce the importance of being looped, a looped heat pipe, TL, was made out of the same material. TL performed to the same maximum attainable power as TU, though this occurred at a different fill ratio; in fact it occurred at a fill ratio very similar to the optimal performance of RL, whereas the un-looped pipe, TU, had an optimal performance at the lower end of the fill ratio spectrum.

When the fluid was changed from de-ionized water to ethanol a remarkable change was seen in the operation of the triangular channel heat pipes TL and TU. The un-looped heat pipes show a maximum attainable power occurring at the lowest fill ratios, while the looped version still has its maximum attainable power occurring at higher fill ratios of approximately 30%. This reveals a tentative link between the fill ratio and looped/un-looped structure which must be examined further. On a side note, ethanol tests were not carried out in the rectangular looped heat pipe as there were concerns the ethanol would negatively react with the plastic end-cap.

Overall, the smaller channel heat pipes performed significantly better than the larger channel heat pipes with ethanol. An explanation for the poor performance of water in the smaller channel heat pipes compared to the large channel looped heat pipe (RL) may have to do with the physical property differences of the fluids, especially the surface tension. Further testing with these and other fluids will help identify the exact cause.

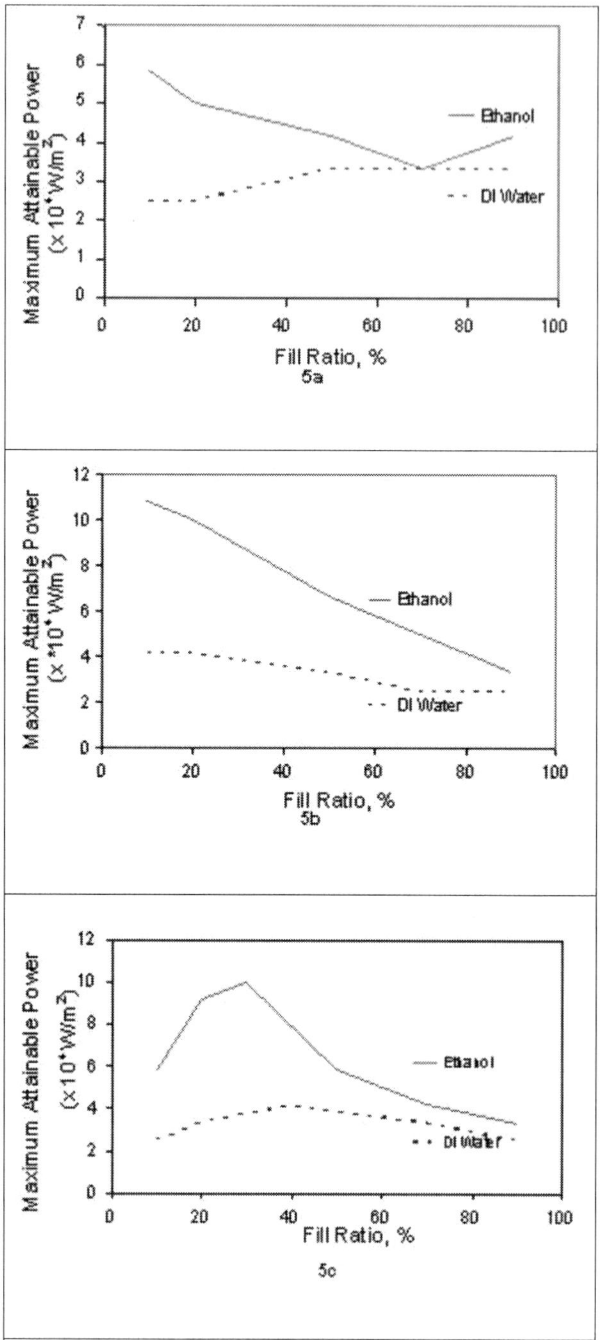

Figure 5 – Maximum attainable power levels at various fill ratios and using different fluid mediums for (a) RU (b) TU and (c) TL.

FUTURE EXPERIMENTATION

One minor change that should bring about increased performance is to reduce the thermal resistance between the heat source/heat sink and the heat pipe. Currently the temperature drop across the interface of the heat source and heat pipe is very high; in the tens of degrees.

The next round of experimentation will use a high heat flux heat source to mimic the power levels of an electronic component. Not only must the heat pipe function well at a higher power, but must be able to handle spikes in power that go significantly higher than the average operational power setting of an electronic device [8]. Hot spot heating is another challenge of cooling electronic devices that will be tested using the new heater. In order to perform experiments at higher power settings, a new thermal interface material will be required as the temperature difference across this material becomes a limiting feature in high power applications [4]. Another limiting feature is the method of heat removal from the condensation region. Currently an aluminum extruded fin heat sink is used. At higher power settings a pin heat sink will be used. These heat sinks have improved efficiency over a fin sink due to their greater surface area, and allowance for 3D air flow which reduces pressure drops and turbulence [4].

CONCLUSION

In summary, optimal performance of a pulsating heat pipe greatly depends on the physical characteristics of the heat pipe such as the number of channels, channel size, and looped/un-looped configuration, as well as on the fluid characteristics such as the type of fluid and the fill ratio. It is important to understand how these parameters affect the operation of a pulsating heat pipe to aid in the design of these heat pipes for realistic cooling solutions. This study thus far reveals the importance of a large number of turns when designing an un-looped heat pipe, and that the optimal fill ratio is higher for the looped channel than the un-looped channel configuration.

ACKNOWLEDGEMENTS

The authors would like to thank the Centre for Microelectronics Assembly and Packaging (CMAP) and Materials and Manufacturing Ontario (MMO) for financially supporting this work. In- kind contributions of Celestica (a heater block) and Brazeway (extruded multi-port tubing) are also gratefully acknowledged.

References

1 G. Moore, Cramming more components onto integrated circuits, **Electronics** 38 (1965)

2 Intel Corporation Website Information, Accessed: March 3, 2005. http://www.intel.com/technology/silicon/mooreslaw/index.htm

3 G. P. Peterson, H. B. Ma, Temperature Response of Heat Transport in a Micro Heat Pipe, **Journal of Heat Transfer** 121 (1999) 438 – 445

4 R. J. McGlen, R. Jachuck, S. Lin, Integrated thermal management techniques for high power electronic devices, **Applied Thermal Engineering** 24 (2004) 1143 - 1156

5 S. Nishio, Oscillatory- Flow Heat Transport Device, 11th International Heat Pipe Conference Preprint Vol. 3, September 12-16, 1999, Japan

6 J.J. Gu, M. Kawaji, R. Futamata, Effects of Gravity on the Performance of Pulsating Heat Pipes. **Journal of Thermophysics and Heat Tranfser**, 18 (3) (2004) 370-378

7 M. B. Shafi, A. Faghri, Y. Zhang, Analysis of heat transfer in un-looped and looped pulsating heat pipes, **Int. J. Numerical Methods for Heat and Fluid Flow** 12 (2002) 585 – 609

8 R. Downing, G. Kajasoy, Single and two-phase pressure drop characteristics in miniature helical channels, **Experimental Thermal and Fluid Science** 26 (2002) 535-546

121e Numerical Study on the Gas and Liquid Slugs for Taylor Flow in a Microchannel

Adeniyi Lawal, Dongying Qian

The extremely large surface-to-volume ratio and the short transport path in microchannel-based reactors enhance heat and mass transfer dramatically and hence provide many potential opportunities for chemical process development and intensification. However, the development of microreactor technology raises new technical challenges that need to be addressed for successful application in chemical reaction engineering. For gas-liquid two-phase flows in microchannels dominated by surface tension effects, the Taylor slug flow regime is the most commonly encountered flow pattern. The present study focuses on the development of the gas and liquid slugs for such flow patterns in a microchannel, since the hydrodynamic characterization is very important for understanding the transport phenomena. In this study, a T-junction reactor (either as packed-bed or with catalytic wall) of varying width; 0.25, 0.5, 0.75, 1, 2 and 3 mm, served as the model micro-reactor, and a finite volume based commercial CFD package, Fluent, was adopted for the numerical simulation of the flow. The slug length, slug velocity, and velocity profile for the gas and liquid slugs as well as the reactor pressure drop were obtained for various operating conditions. The simulation results were in good agreement with our experimental results, and reported literature data. Several correlations for the predictions of the flow characteristics were developed based on the simulation results.

122a Conversion Optimization for Geothermal Power Plants

Gerald R Nix

Geothermal power plants represent an engineering challenge because of lower source temperatures compared to fossil fired power plants. The resulting low conversion efficiency represents a challenging opportunity to the engineer to optimize the cost of produced electricity. There are practical constraints such as the conversion cycle, the type of cooling used in the turbine exhaust condenser, the use of recuperative heat recovery, integration of the power plant with the resource, and others that must be considered. This paper explores the options for optimization of the conversion plant and provides practical guidelines for the geothermal power plant designer.

122b Pyrolysis of Alfalfa Stems and Dedicated Herbaceous Energy Crops in the National Biomass Initiative Program

Akwasi A Boateng

In order to tackle the challenges of developing a biobased industry in the U.S., the 2000 biomass research and development act directs the US-DOE and USDA to integrate technology R&D programs through the Biomass Initiative. Current conversion technologies under the initiative are based on the sugar and the thermochemical platform technologies. Between these two platforms thermochemical conversion shows nearer-time promise towards developing integrated bio-refineries in the US. Pyrolysis is one process under this platform that has gained attention because it can be used to produce pyrolytic oil intermediates that can be subsequently refined to form valuable chemicals and energy carriers such as hydrogen. The role which the choice of biomass plays in achieving economic competitiveness compared with fossil fuel conversion is well documented and hence research emphasis has been placed on dedicated energy crops. In this submission, the results of the pyrolysis of alfalfa stems and certain herbaceous crops specifically, switchgrass and reed canary grass, which are identified under the biomass initiative program as dedicated energy crops, are presented. The pyrolysis was carried out in a PY-GC/MS system to determine the effect of temperature and crop maturity on gas yield. The results show distinct yield patterns and differences between the alfalfa and the herbaceous forages. The information is useful for the evaluation and comparison of the various conversion routes identified under the initiative.

122c Concentrated Acid Pretreatment for the Conversion of Lignocellulosic Materials to Sugar

Roger D Hester, William S Miller

The first stage of a two-step concentrated sulfuric acid process that converts softwood sawdust to sugars has been explored. The research focuses on the ability of an in-house custom fabricated co-rotating twin-screw reactor (TSR) to effectively breakdown and solubilize crystalline cellulose into low molecular weight carbohydrates. This is achieved through intense mechanical shearing action of the lignocellulosic solids during exposure to sulfuric acid at elevated temperatures in the TSR. After the TSR system was placed into operation, an initial two-level, three-factor screening design of experiments (DOE) was performed to determine the importance of screw speed (70 versus 110 rpm), reactor temperature (30 versus 50 ^oC) and sawdust feed rate (2 versus 3 g/min). This screening DOE showed that screw speed was relatively insignificant in comparison to sawdust feed rate and reactor temperature. Based on the DOE screening results, a four-level, two-factor experimental model building DOE was undertaken. In this design two independent variables, sawdust feed rate and TSR temperature, were varied from 2 to 5 g/min and 40 to 70 ^oC. Solid sawdust conversion to liquid, screw torque and TSR exit pressure were measured or recorded to yield percent conversion of solids, material energy requirements and processed material viscosity. Thereafter, model quadratic equations were fitted to the experimental data and found to be statistically significant.

122d Feedstock Flexible Fluidizable Reforming Catalyst Development

Esteban Chornet, Kimberly A Magrini, Richard French, Stefan Czernik, Yves Parent

Commercial naphtha reforming catalysts cannot withstand the harsh reaction conditions used in a fluidized bed for producing hydrogen or tailored syn gas from biomass pyrolysis liquids and vapors. Fluidization is required to optimize the contact of catalyst with the feedstock, which tends to form coke instead of being reformed in a fixed bed reactor. Commercial catalysts, which generally consist of nickel coated onto alumina supports, attrit significantly when fluidized with consistent losses of 10-20 wt% per day. Because process economics cannot tolerate such catalyst losses, we needed to identify and/or develop an attrition-resistant support that could withstand fluidized reforming. Numerous support screening experiments showed that a commercially available alumina material produced by a ceramics manufacturer could have the necessary strength to withstand fluidizing conditions. Catalysts made from these supports have been used to reform pyrolysis liquids and vapors to hydrogen and syn gas and the generated performance data used to refine catalyst compositions. We will report the results of these catalyst performance tests and describe the application of a multivariate technique to the development of catalyst compositions for refoming renewable feedstocks to fuel gases.

122e Solar-thermal Decarbonization of Natural Gas – A Near Term Process

Alan Weimer, Allan Lewandowski, Raymond Hobbs

The production of hydrogen and amorphous structured carbon black has been demonstrated at high rates using a solar-thermal fluid-wall transport tube reactor process. Experimental results will be summarized. In order to assess the near-term viability of the process, design and economic considerations are given to the processing of a slip stream being fed to a natural gas power plant in the Phoenix, AZ area. The reactor is heated to 1500 C using three heliostat fields, each focusing sunlight on one of three secondary concentrators along the length of the reaction tube. Carbon black is collected in a downstream baghouse filter and hydrogen is purified using a pressure swing adsorber (PSA). Some of the purified hydrogen is recycled to provide the protective fluid-wall gas. The process gas stream out of the PSA is primarily methane and hydrogen. The carbon black produced can be sold to the tire industry while the high purity hydrogen can be used to drive a stationary fuel cell. In addition, significant heat recovery from the reactor can be used to produce steam to drive a turbine to generate electrical power. The product gas from the PSA is sent to the NG plant for combustion. Calculations are made to determine the CO2-equivalent avoidance of the plant and the value of the CO2-equivalent avoidance credits in order for the plant to be profitable. It is anticipated that such a process has near term feasibility once greenhouse gas avoidance credits are marketed in the United States.

123a New Opportunities for Thermal Energy Saving in BPA Chemical Plants Using Real-Time Energy Integration

Abd ElRazak El-Saleh, Kamrul Hasan, Mahmoud Bahy M Noureldin

The worldwide increase in energy demands and the uncertainty of future energy costs, make energy conservation a major element in controlling operating costs of chemical processes.

Bisphenol-A (BPA) is an important chemical intermediate used mainly in the manufacture of polycarbonates and epoxy resins. Polycarbonates are both the largest and the fastest growing market for Bisphenol-A, accounting for about 60% of the world demand. The second largest end user (about 30%) for Bisphenol-A is epoxy resins. Approximately 10% of BPA consumption goes into various specialty applications, which also show solid growth from their relatively small bases. They are flame-retardants, unsaturated polyester resins, polysulfone resins, polyacrylates, and so on.

BPA is produced via acetone reaction with phenol in a catalytic stripping reactor utilizing ion exchange resin as catalyst.

Energy usage in BPA production represents a major cost center and process owners must constantly monitor its consumption to enhance their profitability.

Such needs are the motivation for more aggressive energy recovery schemes. Real- time operation optimization of chemical plants can be the answer for more energy recovery in BPA plants. This paper addresses this objective through the application of new user friendly software for real-time thermal energy targeting and the determination of optimal operating conditions that render desired thermal energy consumption objective in BPA plants. This novel software for real-time energy integration is used to investigate the potential benefits of energy savings that can be obtained from implementing real-time operation optimization in a typical BPA chemical plant. The industrial case study introduced in this paper is modified for the sake of confidentiality. However, it demonstrates that moderate energy benefits can be obtained upon applying real-time operations optimization for energy saving. It shows that for typical BPA plant, operational modifications can result in about 15 % savings in utilities. It also shows that simple control of few operating conditions can result in such saving.

These results would support the trend of using cost effective real time energy integration system for energy recovery application in chemical plants.

Energy Optimization Tools for Non-Core Business Applications

Frederick P. Fendt, P.E., Rohm and Haas Company
Glenn Hahn, Spirax Sarco, Inc.
Dr. Anthony Wright, Oak Ridge National Laboratory

1. Introduction

BestPractices, a program of the U.S. Department of Energy's (DOE) Industrial Technologies Program (ITP), works with industry to identify plant-wide opportunities for energy savings and process efficiency. Goals of the BestPractices program include helping companies start to save energy and money, to reduce pollution and emissions, and to increase productivity right now. Through the implementation of new technologies and systems improvements, companies across the United States are achieving immediate results.

As a part of the BestPractices efforts, a suite of tools has been developed to assist industrial plant personnel to identify potentially significant energy efficiency improvements to their plant utility systems. This paper discusses the tools that are available, provides an example of how these tools can be used to develop internal utility system metrics, and discusses how these tools are being used strategically to leverage lean corporate energy efficiency staff throughout a corporation's manufacturing base.

Told from the points of view of an energy systems end user, an energy systems service provider, and a DOE tools developer, this paper provides a 360° look at this valuable suite of tools.

2. U. S. DOE BestPractices Opportunity Assessment Tools

The BestPractices software tools that are presently available are listed and briefly described in Table 1. These software tools share some key characteristics that are noted below:

a. All of the BestPractices software tools focus on energy efficiency opportunity assessment. Opportunity assessment means quantifying the potential savings that is available for a specific improvement opportunity. For example, in steam systems this might be savings that can result from increasing condensate return, for process heating this could be savings that could result from reducing excess oxygen in furnaces. These tools emphasize the major energy savings opportunities that are typically possible in plant utility systems.

b. The BestPractices software tools focus on the major improvement opportunities that are typically available in plant utility systems - those with the biggest bang for the buck. The tools are not designed to identify and quantify every energy savings opportunity that can be available in plant utility systems.

c. BestPractices has partnered with key trade associations and focused technical groups to develop the software tools. Examples include partnership with the Hydraulics Institute to develop the pump tool, partnership with the International Heating Equipment Association to develop the process heating tool, and partnership with the Air Movement and Controls Association to develop the fan tool.

d. All of the BestPractices software tools (except ASDMaster) are available free of charge; the tools can be downloaded from the BestPractices web site - www.eere.energy.gov/industry or www.oit.doe.gov/bestpractices.

e. Experience (documented case studies, etc.) has shown that application of the BestPractices software tools can result in identification of large potential energy and cost savings. Identified savings potentials for individual plant utility systems can often be 10% or greater of the utility system energy bill.

To assist industrial plant personnel and utility system service providers in applying the BestPractices software tools, the BestPractices effort has developed two types of training:

a. End user training – Typically 1 day in length; focuses on reviewing key savings opportunities available for individual plant utilities; end user training also introduces specific BestPractices software tools. Presently, BestPractices presents end user training for motors, compressed air, pump systems, steam systems, process heating systems, and fan systems.

b. Specialist qualification training - Typically 2 to 2-1/2 days in length; specifically focuses on how to effectively use individual BestPractices software tools. At the end of this training the trainees take an exam, and if they pass the exam they obtain the designation of Qualified Specialist for a specific software tool. Specialist Qualification training is presently available for the compressed air, pump, steam, and process heating software tools, and will be available soon for the fan tool.

Table 1. Overview of BestPractices Opportunity Assessment Software Tools

TOOL	TOOL DESCRIPTION
MotorMaster+	Identifies inefficient or oversized facility motors. Computes energy and demand savings associated with selection of replacement energy-efficient models. Includes motor inventory management, maintenance logging, and a motor price and performance database.
Pumping System Assessment Tool (PSAT)	Assesses efficiency of pumping system operations. Uses achievable pump performance data from Hydraulic Institute standards and motor performance data from MotorMaster+ database to calculate potential energy and associated cost savings.
AirMaster+	Assesses compressed air systems. Models existing / future system upgrades. Evaluates savings / effectiveness of energy efficiency measures.
Steam System Scoping Tool (SSST)	Evaluates a facility's steam system operations and management practices against best practices.
Steam System Assessment Tool (SSAT)	Estimates impacts of key steam system improvements. Provides energy, cost, and emission savings results for up to sixteen different steam improvement opportunities.
3E Plus	Calculates the most economical thickness of industrial insulation for user input operating conditions. Performs calculations using the built-in thermal performance relationships of generic insulation materials or user-supplied conductivity data for other materials.
Process Heating Assessment and Survey Tool (PHAST)	Surveys process heating equipment. Identifies most energy-intensive equipment. Performs energy (heat) balances on selected equipment (furnaces) to identify/reduce non-productive energy use. Compares performance of furnaces under various operating conditions and tests "what if" scenarios.
Fan System Assessment Tool (FSAT)	Calculates amount of energy used by fan systems. Determines fan system efficiencies. Quantifies savings potentials of upgraded fan systems.
NOx and Energy Assessment Tool (NxEAT)	Assesses/analyzes NOx emissions and applications of energy efficiency improvements at petroleum refining and chemical plants. Inventories emission from NOx generating equipment. Compares various technology applications and efficiency measures that affect overall costs and reduction of NOx. Performs "what if" analyses to optimize and select the most cost-effective methods of reducing NOx.
ASDMaster	Determines economic feasibility of adjustable speed drive (ASD) applications. Predicts electrical energy savings by using ASDs. Available by contacting Electric Power Research Institute (EPRI).

3. Example - Using BestPractices Tools to Quantify Utility System Metrics

In addition to generating energy improvement recommendations, all the DOE BestPractice Tools provide a means to benchmark and establish good baseline operating metrics for steam, compressed air, process heating, motors, fans, pumps, adjustable speed drives and NOx emissions.

For this paper, the Steam System Assessment Tool (SSAT) is used as an example. For this example, a system with only one boiler will be evaluated (assuming that for a system with multiple boilers, all of the boilers respond similarly to changes in system conditions). Analyses for more complex boiler operations would need to be based on the steam system "impact" boiler and are beyond the scope of this example.

SSAT is a structured software-based steam system simulation model. It is designed to allow the user to build an approximate model of their steam system (maximum of three different steam headers) and to predict savings (cost and emissions) achieved from implementation of key steam system operation projects. SSAT does not offer an exact modeling and simulation solution to match precisely what is happening at a plant, but will nominally produce results in the 80% to 90% range depending on the validity of input data. The goal of SSAT is to provide relatively quick guidance to engineers on how to improve their steam systems.

The SSAT input template contains default data to provide examples of potential input values. A good practice is to print a copy of the input sheet to gather the required information. The more accurately the input data correlates with known operating conditions, the closer users can get to a true benchmark or baseline metrics of their site total steam systems.

Examples of input data include:

- Site power import; site power cost (past 24 months of costs, $/kWh)
- Boiler pressure; boiler DA tank pressure (past 12 months)
- Boiler fuel; boiler efficiency; make-up water costs (past 24 months of costs and efficiency)
- Steam turbine(s) (type in use and efficiency)
- HP/MP/LP pressures; HP/MP/LP steam end flows (past 12 months, PPH)
- Condensate return or boiler feedwater flows (past 12 months, GPM)
- Number of HP/MP/LP failed open steam traps (steam trap survey results)
- Number of HP/MP/LP steam leaks and insulation efficiency (steam leak, insulation survey results)

The SSAT *Model* worksheet (Fig. 1) then displays a schematic representation of the simulation of the current steam system operation as specified from the collected input data.

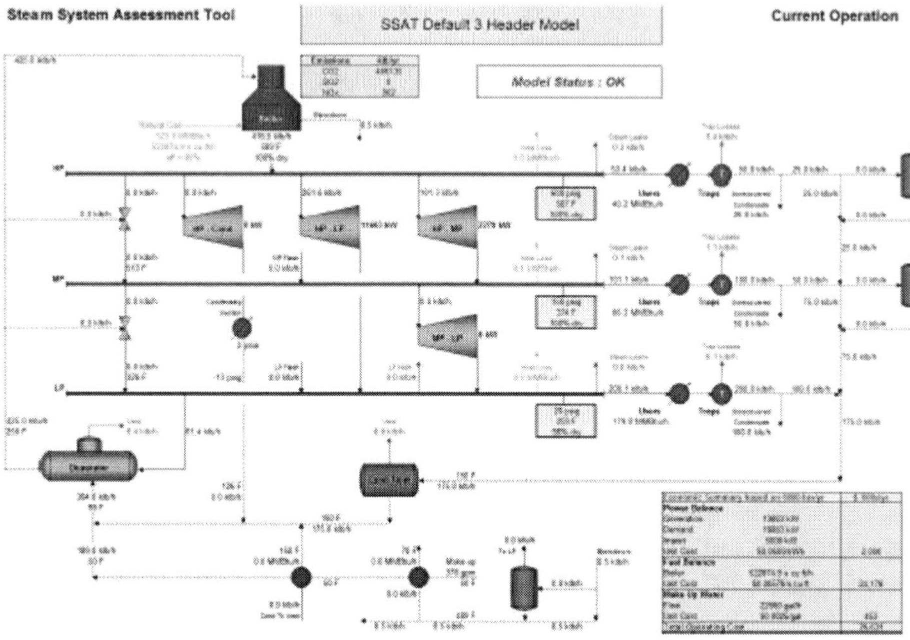

Figure 1. Example of SSAT Software Model Worksheet

Users should refer to the *Model* worksheet in order to check and re-validate that the system is being simulated as expected (compare *Model* "Total Operating Cost" to actual total operating cost. If not, adjustments should be made accordingly (e.g., change in HP/MP/LP flows, condensate recovery percentage, steam trap and steam leak losses).

When the SSAT *Model* "Total Operating Cost" begins to match the actual total operating cost, a true benchmark or baseline metric steam system model of current operation has been achieved (note that this is true only for systems that have one boiler or multiple boilers that all respond similarly to changes in steam system changes). Now a number of checkpoints and steam system project improvement opportunities can be determined, reviewed, and validated by using the *Project* input sheet.

Examples:

1. Present "Marginal Cost of Steam" (cost of supplying an additional 1 klb of steam at the inputed pressure levels) can be determined for trending purposes.
2. Steam measurement check to existing steam meters or the importance of actual steam measurement versus assumed incorrect steam flows when no meters are used.
3. Alternative boiler fuels; appropriate time to change.
4. Improved boiler efficiency.
5. Steam demand reductions or steam management.
6. Improved boiler blowdown and heat recovery.
7. Options in adding or removing steam turbines.
8. Improved condensate return.
9. Addition of flash or heat recovery.
10. Reduction in steam losses from leaks, failed steam traps, insulation, etc.

The SSAT *Results* worksheet shows the key results (steam energy costs and emission changes) from the models of both the current operation and the system operation after the implementation of chosen project opportunities. The results are presented in a tabular form, allowing a quick assessment of the project impact.

The chosen project savings from the benchmark or baseline metric developed from the DOE SSAT modeling program can now be used to perform a "Capital Project" evaluation. A simple "Capital Project Decision Model," as illustrated in Fig. 2 would use the following input data:

1. Steam & Condensate Improvement Project
2. Steam system energy savings for condensate and steam loss improvements ($553,000/yr)
3. Impact years for the project (5)
4. Project equipment costs ($95,255.50)
5. Project installation costs ($46,401.60)
6. Freight Costs ($1,500.00)

CAPITAL PROJECT DECISION MODEL

CUSTOMER: AIChe Capital Project Example

PROJECT: Steam & Condensate Improvement Project)
Consists of: 1. Replacing failed open steam traps.
2. Replacing failed condensate pumps.
3. Repair steam leaks.

Investment Outlay at t=0	
Project Equipment	$95,255.50
Ancillary Equipment	$0.00
Installation	$46,401.60
Freight	$1,500.00
Total Investment Outlay	$143,157.10

Cost Savings	
Impact Years	5
Cost Savings	553,000

Depreciation Expense		Year									
Basis	$143,157.10										
Life	5	1	2	3	4	5	6	7	8	9	10
Depreciation Expense		$28,631.42	$28,631.42	$28,631.42	$28,631.42	$28,631.42	$0.00	$0.00	$0.00	$0.00	$0.00

PROJECT CASH FLOW ANALYSIS

	Year 0	Year 1	Year 2	Year 3	Year 4	Year 5	Year 6	Year 7	Year 8	Year 9	Year 10
Net Investment Outlay	($143,157.10)										
Cost Savings		$553,000.00	$553,000.00	$553,000.00	$553,000.00	$553,000.00	$0.00	$0.00	$0.00	$0.00	$0.00
Less: Depreciation		$28,631.42	$28,631.42	$28,631.42	$28,631.42	$28,631.42	$0.00	$0.00	$0.00	$0.00	$0.00
Before Tax Savings		$524,368.58	$524,368.58	$524,368.58	$524,368.58	$524,368.58	$0.00	$0.00	$0.00	$0.00	$0.00
Taxes @ 40%		($209,747.43)	($209,747.43)	($209,747.43)	($209,747.43)	($209,747.43)	$0.00	$0.00	$0.00	$0.00	$0.00
Net Savings		$314,621.15	$314,621.15	$314,621.15	$314,621.15	$314,621.15	$0.00	$0.00	$0.00	$0.00	$0.00
Add Back Depreciation		$28,631.42	$28,631.42	$28,631.42	$28,631.42	$28,631.42	$0.00	$0.00	$0.00	$0.00	$0.00
Net Operating Cash Flow	($143,157.10)	$343,252.57	$343,252.57	$343,252.57	$343,252.57	$343,252.57	$0.00	$0.00	$0.00	$0.00	$0.00

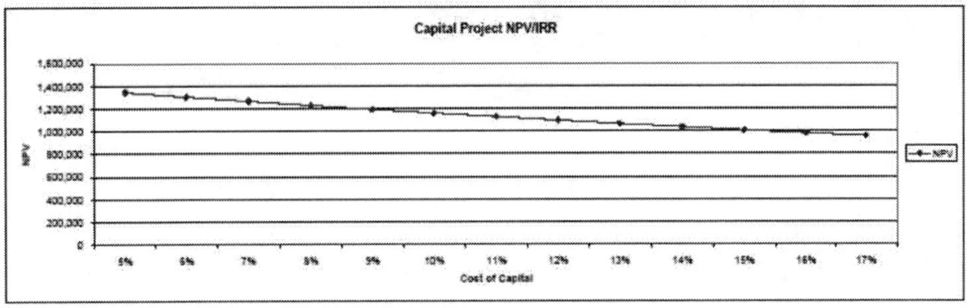

Cost of Capital	NPV
5%	1,342,947
6%	1,302,748
7%	1,264,246
8%	1,227,351
9%	1,191,976
10%	1,158,040
11%	1,125,469
12%	1,094,192
13%	1,064,142
14%	1,035,257
15%	1,007,479
16%	980,753
17%	955,027

	PROJECT
IRR	239.2%
PAYBACK MO.	5.0

Return on Investment	PROJECT
	219.9%

Figure 2. Capital Projects Decision Model Results, Based on SSAT Calculations

The proposed Steam & Condensate Improvement Project as developed from the SSAT software project opportunity results shows:

1. Internal rate of return (IRR) (239%)
2. Payback period required to recover the original project investment (5 months)
3. Return on investment (220%)

4. Effectively Implementing BestPractices Tools in Chemical Plants

The U.S. chemical industry has undergone significant and profound changes in the last twenty years and promises to continue that trend. In the late 1980s and early 1990s, the trend in the U.S. chemical industry was to re-engineer the workplace. In other words, the strategy was to add more value with fewer people. The American Chemical Society has calculated that the major U.S. chemical firms went from producing $221,600/employee in 1991 to $350,700/employee in 2001.[1] By the mid-1990s, the emphasis had shifted to a focus on growth. Companies were still reorganizing, but were eliminating functions that were not highly focused on their core businesses. Both trends still continue today. For example, for engineers, the Bureau of Labor Statistics of the U.S. Department of Labor estimates that in the ten years between 2002 and 2012, there will be 21% fewer engineers working in the non-pharmaceutical chemical industry in the United States.[2]

The net result of this trend is that, for non-core business disciplines, there are fewer practitioners today in the U.S. chemical industry than there were in the past. For example, few plants have dedicated utility system engineers. Even at the corporate level, usually only the larger and more progressive companies have engineers responsible for energy and utility system operating efficiency, and these are usually a staff of only a very few engineers for the entire company's manufacturing operation.

However, this reduction in the number of energy system expert practitioners has come at a time when the U.S chemical industry is faced with historically unusually high prices for energy and fuels (Fig. 3).

1 C&EN June 24, 2002 http://pubs.acs.org.cen
2 "Total Employment by Occupation and Industry, 2002 and Projected 2012"

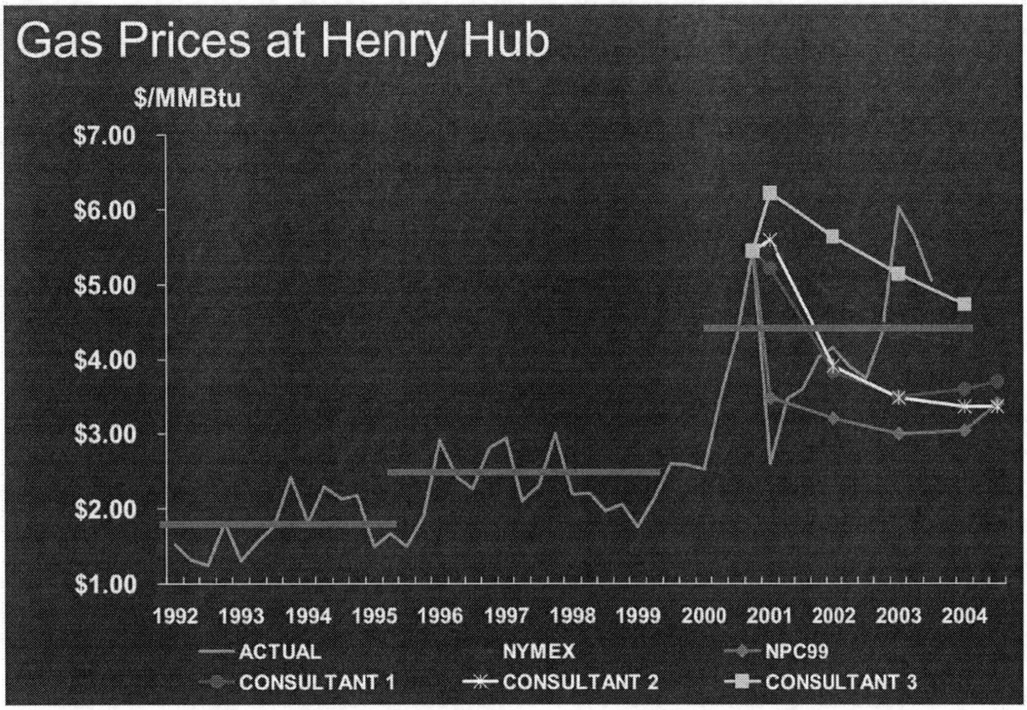

Figure 3. Recent History of "Henry Hub" Gas Prices

Energy, maintenance, and labor are the hightest of the controllable costs in chemical manufacturing. However, with no dedicated energy engineers at most plants, and only a very few, if any, at the corporate level, assessment tools, such as the U.S. DOE tools, can be a critical part of finding the significant cost saving opportunities that exist in the manufacturing facilities.

For example, Rohm and Haas engineers used the tools to help their plants identify energy efficiency opportunities and to screen potential efficiency improvement projects. This can be done with the help of corporate resources, but the tools are also available to all plant engineers to use without corporate assistance.

The Rohm and Haas Company is an inventor of specialty materials that improve the quality and functionality of everything from semiconductor chips to house paint. As a technology company that prides itself on delivering technically advanced products and services to its customers, it makes sense that Rohm and Haas is also committed to reducing the energy intensity of its own manufacturing plants wherever it is economically feasible.

Ideally, the corporate energy efficiency expert practitioners could visit every one of Rohm and Haas' sites and personally identify these opportunities and then help make sure that they were implemented. Additionally, in an ideal world, they could periodically revisit each of the sites to make sure that energy efficiency practices were continuing to receive appropriate attention, and to identify any new opportunities that may have developed because of plant expansions, changing infrastructure, or changing fuel costs, as examples.

However, with only a handful of energy efficiency staff and over 100 manufacturing and research facilities located in 25 countries, the reality is that corporate staff can only visit a handful of sites in a given year. However, the availability of the DOE tools allows Rohm and Haas to find and implement many more opportunities than might otherwise be practical.

At Rohm and Haas, the tools are introduced through informal brown-bag lunch time sessions, as well as on site training when requested. Tools such as WebEx allow sites around the world to participate for almost no cost in these informal sessions. Support is also provided through the DOE's end user training when it is offered in locations convenient to Rohm and Haas' manufacturing sites.

At the corporate level, many of the successes from using the tools never get reported as such. The plant level user may get an idea, or validate an idea, or perhaps quantify the value of an idea, from using the tools. He or she would then implement the idea and appreciate the savings, all without the involvement, or even perhaps without the awareness of anyone at the corporate level. This is, of course, one of the goals of the tool deployment strategy; i.e., to encourage the plant level engineers to improve their energy efficiency. However, some of the successes are known.

Rohm and Haas' Deer Park plant is located just outside of Houston, Texas. In 1998, plant engineers wanted to increase the hydraulic pumping efficiency of one of the plant's cooling tower systems. As a part of this effort, the DOE performed a pump assessment in using the Pumping System Assessment Tool (PSAT). The system investigated contained four single stage vertical turbine pumps, with three pumps online during normal operation. Each pump had a design capacity of 25,000 gpm at 120 ft. TDH and a pump curve efficiency of 80%. Measured efficiencies were found to be much lower.

Using information from PSAT, plant engineers proposed a new pumping system that improved capacity and reduced energy usage. The new pumping system achieves between 87% and 88% efficiency in the range of 23,000 and 27,000 gpm and a TDH between 120 and 130 ft. Said simply, the new system delivers more flow capacity while using less energy. The electrical energy savings alone totals over $70,000 annually for the system when operated as a three-pump system. The increased capacity allows the operators to run only two pumps for much of the year, significantly increasing the savings.

5. Summary and Conclusions

The case studies in this paper illustrate the value of the U.S. DOE BestPractices opportunity assessment software tools to the chemical industry. The tools can be applied to identify important metrics for energy efficiency improvements in chemical plants. They can also be used to assist chemical plant personnel to strategically identify and implement energy efficiency improvements in their plants. For more information on tools, resources, and training to assist chemical plant personnel to effectively use the BestPractices tools, visit the www.eere.energy.gov/industry or www.oit.doe.gov/bestpractices web sites.

123c Innovative Absorber/Stripper Configurations for CO_2 Capture by Aqueous Monoethanolamine

Gary T Rochelle, Majeed S Jassim

The up-to-date technology to capture CO_2 from coal-fired power plants is absorption/stripping with aqueous monoethanolamine (MEA). The energy consumption in stripping can be 15 to 30% of the power plant output. A rigorous rate-based model was used to simulate several flowsheet alternatives that reduce the energy requirement using ASPEN simulation environment. The scenarios considered are: 1. Simple stripper with 90% CO_2 removal, 2. Simple stripper with 95% CO_2 removal, 3. Stripper with vapor recompression with 90% CO_2 removal, 4. Multi-pressure stripper with vapor recompression with 90% CO_2 removal, 5. Multi-pressure stripper with cooling water inter-cooling on compressor stages with 90% CO_2 removal. The potential saving as "equivalent work of steam/mole CO_2 removed" is 25-40% using these innovative design scenarios in comparison to conventional MEA stripper.

Automated Design and Optimization of Flexible Utility Systems under Variable Operating Conditions

Oscar Aguilar, Robin Smith, Simon Perry, and Jin-Kuk Kim
Centre for Process Integration, The University of Manchester, UK
PO Box 88 Sackville Street, Manchester, M60 1QD, UK

Prepared for Presentation at the 2005 Spring National Meeting, April 2005,
"Energy Tools and Methodologies for Industrial Applications"
Unpublished

ABSTRACT

Heating, cooling, and power needs of most process industries are satisfied by in-site utility plants supplying steam, electricity and rotational power. These systems are typically comprised of many units of various types that can be interconnected in different ways, offering a wide number of design and operational options. Although it is possible to achieve substantial savings by selecting the alternatives that minimize operating and/or capital costs, such procedure represents a large combinatorial problem that requires complex calculations involving the performance of all units. Moreover, the task becomes even larger when the plant operates under variable conditions since it is necessary to consider several scenarios at the same time and establish the appropriate equipment sizes and loads for a flexible operation of the system. Thus, the present work proposes a novel modelling framework for utility plants that allows structural and operational variables to be optimized simultaneously in order to fully exploit the flexibility of these systems when subject to fluctuating conditions. Furthermore, whenever failure and maintenance situations are evaluated, the procedure will determine the optimal equipment redundancy to ensure reliable operation under such circumstances. Finally, due to its inherent robustness, the suggested (MILP) formulation can tackle effectively real industry cases including operational optimisation, grassroots and retrofit designs.

1. INTRODUCTION

In order to meet the energy demands from industrial processes, utility plants must convert large amounts of basic feeds (e.g. fuel and water) into steam, electricity, and rotational power. Usually many interconnected units like boilers, steam turbines, and gas turbines must constitute a flexible and reliable system capable of fulfilling the variable heat, cooling, and power requirements under different conditions (e.g. energy prices, ambient temperature). Therefore, the design and operation of utility systems imply an extensive number of alternatives that, if properly selected, can yield considerable savings in operating and capital costs. However, to minimize such expenditure it is necessary to assume a final configuration of the plant, determine the actual performance of each piece of equipment, simulate the operation of the whole system, and establish the overall economic consequences of every option. Because of the highly combinatorial nature of such task and the complex calculations involved, conventional approaches to these problems have addressed either the operation or the design (synthesis) of utility systems separately.

On one side, operational problems apply to existing utility plants in which structural modifications are not contemplated. In such cases the objective is to establish the operating conditions for all flowsheet streams (e.g. temperature, pressure, mass flow) and equipment units (e.g. on-off status and load) that will minimise costs while satisfying the specified energy/water demands. Since normally there are numerous possibilities to fulfil such requirements, mathematical optimisation techniques have been widely applied in order to evaluate and compare these alternatives more quickly and reliably through the use of computational algorithms. For instance, Nath and Holliday (1985) reported the implementation of a MILP software for the optimisation of utility systems in several processing plants. Iyer and Grossmann (1997) extended this concept to multiperiod operation including transition costs due to load changes and switching off/on of the units. More recently, Varbanov et. al. (2004) proposed a successive-MILP procedure that enables the utilization of more accurate equipment models.

On the other hand, for design (synthesis) cases the final configuration of the utility plant must be established given that some investment is available either to erect the system from scratch or to revamp an existing one. Apart from the operational issues, in such cases it is also necessary to define the types of units to be installed, their number, sizes, and the interconnections among them. However, conventional methods have simplified the design problem assuming that all units operate at full load to satisfy a single set of (constant) demands and conditions. Even with these simplifications the large number of structural alternatives require the use of mathematical optimisation tools to determine the options that minimize capital and operating costs (e.g. Papoulias and Grossmann, 1983 and Bruno et. al. 1998). While these approaches can yield acceptable results for relatively stable cases, in the present context of energy production (e.g. de-regulated electricity markets) it is indispensable to take into account several operating scenarios simultaneously and consider part-load operation of the equipment to avoid large economic penalties.

In response to these incentives, there have been methodologies combining elements of both operational and design problems. For example, Hui and Natori (1996) evaluated several retrofit options along different operating conditions through a MILP model. Maia and Qassim (1997) employed simulated annealing to select the optimal configuration for a utility system subject to variable demands. Maréchal and Kalitventzeff (2003) applied a multiperiod MILP formulation to choose the optimal utility plant design from several choices obtained with targeting techniques. Nevertheless, in all these studies equipment sizes and loads are not optimised simultaneously as continuous functions and/or the user must pre-specify certain design options with fixed-sized units that often miss better alternatives. Conversely, the present work proposes a common optimisation framework for the automated design, retrofit and operation of industrial utility plants under variable operating conditions. The suggested method fully exploits the flexibility of these systems by considering part-load operation for all units within a simultaneous structural and operational optimisation. The next sections of this article will first describe the proposed modelling approach, followed by an overview of the whole methodology, and a case study showing how enhanced solutions can be obtained with this technique.

2. EQUIPMENT MODELLING

Mathematical models are employed to predict the performance of each piece of equipment within the plant and simulate/optimise the whole system. Although there are many types of models and optimisation techniques currently available, linear modelling offers robust and reliable routines suitable for real industry problems. While conventional linear models calculate equipment performance either as a function of load (i.e. fixed unit sizes for existing plants) or as a function of size assuming full-load operation (i.e. equipment at part-load not contemplated), in order to carry out simultaneous structural and operational optimisation it was necessary to develop novel generic models that establish equipment performance as a linear function of both unit size and load.

Figure 1: Non-linear curves on efficiency graphs appear as linear trends when the data is plotted on an output vs. input chart

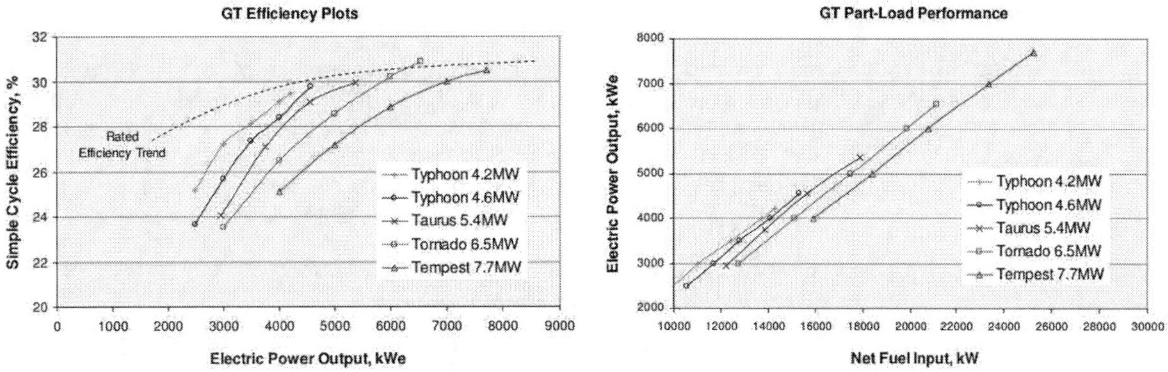

The proposed models can reflect non-linear efficiency trends employing linear equations but without resulting in excessive computational expenditure (i.e. without piece-wise linearization). The key feature to achieve this is to avoid the explicit calculation of the efficiencies during the optimisation as they can always be obtained afterwards if necessary (e.g. dividing power produced over fuel consumed). Figure 1 illustrates the fact that non-linear efficiency curves correspond to linear trends when the information from the first one is presented on an output vs. input plot (i.e. both graphs are perfectly equivalent). Employing this approach it has been possible to derive linear performance models directly from manufacturer's data and achieve very good accuracy for preliminary design purposes (i.e. errors within ±5%). Additionally, if more precision is required, the models can be adjusted to fit the behaviour of specific units as in the case of existing plants. In the next sub-sections the main characteristics of each equipment model will be explained.

2.1. Multifuel Utility Boilers

Within large industrial plants utility boilers are extensively employed to satisfy process heating demands because they offer a convenient way of generating high-pressure steam that can be used to produce in-site power before being delivered to consumers. Boiler performance varies with equipment size, load and the type of fuel (or fuel mixture) being actually consumed. These degrees of freedom must be taken into account for the simulation/optimisation of a utility system and, therefore, the suggested model is capable of assessing such tradeoffs. The first step to derive this model is the linear relationship between fuel consumption and steam production that most boilers follow:

$$Q_f^{boi} = B^{boi} \cdot Q_{stm}^{boi} + C^{boi} = B^{boi}(\Delta h_T^{boi} \cdot M_{stm}^{boi}) + C^{boi} \quad (1)$$

$$M_{stm}^{boi} = \frac{1}{\Delta h_T^{boi}}\left(\frac{Q_f^{boi}}{B^{boi}}\right) - \frac{C^{boi}}{\Delta h_T^{boi} \cdot B^{boi}} \quad (2)$$

Where:

Q_f^{boi} Net heat from the fuel combustion in the boiler (kW)

Q_{stm}^{boi} Net heat added to the steam/water in the boiler (kW)

Δh_T^{boi} Enthalpy difference between feed water and outlet steam conditions (kJ/kg)

M_{stm}^{boi} Current steam mass flow produced in the boiler (kg/s)

B^{boi}, C^{boi} Slope and intercept regression parameters (adim, kW)

When several fuels are accessible in the site, equation (1) is used to obtain the regression parameters corresponding to boiler operation burning individual fuels. Afterwards, equation (2) can be extended to consider the consumption of a fuel mixture inside the boiler:

$$M_{stm}^{boi} = \frac{1}{\Delta h_T^{boi}}\left(\frac{Q_{f1}^{boi}}{B_1^{boi}} + \frac{Q_{f2}^{boi}}{B_2^{boi}} + \ldots + \frac{Q_{fn}^{boi}}{B_n^{boi}}\right) - \frac{C_{avg}^{boi}}{\Delta h_T^{boi} \cdot B_{min}^{boi}} \quad (3)$$

$$\eta_{boi} = \frac{\Delta h_T^{boi} \cdot M_{stm}^{boi}}{Q_{f1} + Q_{f2} + \ldots + Q_{fn}} \times 100 \quad (4)$$

Where:

$Q_{f1}^{boi}, Q_{f2}^{boi} \ldots Q_{fn}^{boi}$ Net heats released by fuels 1,2...n inside the boiler

$B_1^{boi}, B_2^{boi} \ldots B_n^{boi}$ Slope regression parameters for single-fuel operation (adim)

$C_1^{boi}, C_2^{boi} \ldots C_n^{boi}$ Intercept regression parameters for single-fuel operation (adim)

B_{min}^{boi} Lowest slope regression parameter (adim)

C_{avrg}^{boi} Average intercept regression parameter (kW)

η_{boi} Boiler efficiency (%)

Equation (3) establishes boiler production from the actual amounts of different fuels being consumed by the unit and offers the advantage of a linear function that also evaluates the impact of equipment size and load. Note that boiler efficiency might be calculated after optimisation with equation (4) (i.e. a non-linear function). The effect of these variables on boiler performance is illustrated on Fig. 2, which shows how bigger units have higher design efficiencies and are able to better keep this efficiency at part-load operation.

Figure 2: Effect of boiler size, load and type of fuel on boiler performance.

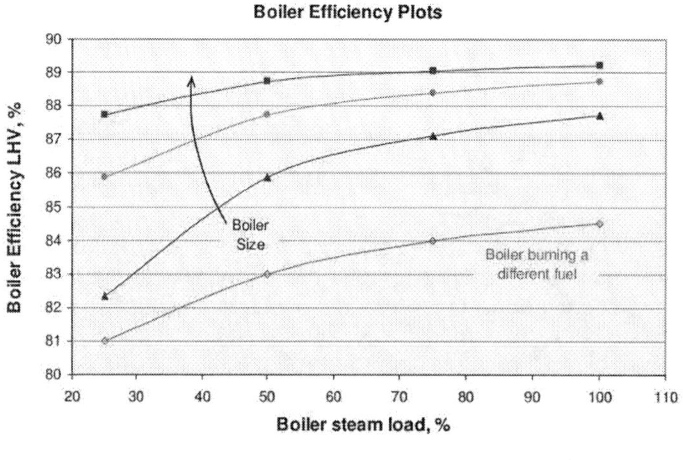

2.2. Steam Turbines

This type of turbines can be found in almost any processing site where steam is accessible, particularly in the chemical, paper, and metal industries. They are employed to expand steam to a lower pressure in order to produce shaft power and drive all kind of rotational equipment. Their performance not only varies with equipment size and load, but also with inlet steam conditions (pressure, temperature) and discharge pressure. In order to reflect such features, the proposed model makes use of an expression for the expected design performance together with the Willans line to account for part-load operation of the units (i.e. equations (5) and (6)):

$$W_{is\,D}^{st} = \Delta h_{is} \cdot M_D^{st} = A^{st} + B^{st} \cdot W_D^{st} \qquad (5)$$

$$W^{st} = n^{st} \cdot M^{st} - W_{int}^{st} \qquad (6)$$

$$n^{st} = \Delta h_{is} \frac{L^{st}+1}{B^{st}} \qquad (7)$$

$$W_{int}^{st} = L^{st} \cdot W_D^{st} + (L^{st}+1)\frac{A^{st}}{B^{st}} \qquad (8)$$

$$W^{st} = \Delta h_{is} \frac{L+1}{B^{st}} M^{st} - L \cdot W_D^{st} - (L+1)\frac{A^{st}}{B^{st}} \qquad (9)$$

Where:

$W_{is\,D}^{st}$	Isentropic design power (kW)	A^{st}, B^{st}	Intercept and slope regression parameters for design operation (kW, adim), both also depend on ΔT_{sat}^{st}
M_D^{st}, M^{st}	Design and actual steam turbine mass flow (kg/s)	L^{st}	Proportion coefficient for power losses (adim), it also depends on ΔT_{sat}^{st}
W_D^{st}, M^{st}	Design and actual steam turbine shaft output (kW)	ΔT_{sat}^{st}	Saturation temperature difference corresponding to turbine inlet and outlet steam pressures
Δh_{is}	Isentropic inlet-outlet enthalpy difference (kJ/kg)		
n^{st}	Willans line slope (kJ/kg)		
W_{int}^{st}	Willans line ordinate intercept (kW)		

Since equations (5) and (6) must be compatible for full-load operation and all regression parameters depend on steam inlet/outlet conditions, it is possible to deduce a final expression (equation (9)) to determine the actual steam turbine output as a linear function of equipment size (design output) and actual steam consumption given certain inlet/outlet steam conditions. Additionally, the actual heat contained in the exhausted steam is also calculated with a linear function due to the fact that apparent power extracted from the steam is proportional to the isentropic power of the turbine:

$$W_{app}^{st} = \alpha^{st} \cdot W_{is}^{st} + \beta^{st} \qquad (10)$$

$$W_{app}^{st} = M^{st}(h_{in} - h_{out}) = M^{st} \cdot h_{in} - Q_{out}^{st} = \alpha^{st} \cdot W_{is}^{st} + \beta^{st} \qquad (11)$$

$$Q_{out}^{st} = M^{st} \cdot h_{in} - \alpha^{st} \cdot W_{is}^{st} - \beta^{st} \qquad (12)$$

Where:

W_{app}^{st}	Apparent power extracted from the steam (kW)	Q_{out}^{st}	Heat contained in the exhausted steam (kW)
h_{out}	Specific enthalpy of the exhausted steam (kJ/kg)	α^{st}, β^{st}	Regression parameters (adim, kW)

Equations (9) and (12) can reproduce the impact of the major variables on turbine performance, for example, with a higher inlet enthalpy or a lower exhaust pressure, the energy potential (i.e. enthalpy difference) increases and the equipment will require less steam for the same output. Also, bigger units can generate more power from the same energy potential and exhibit lower efficiency drops at partial loads (i.e. they are more efficient and reject less heat in the exhaust). Note that, although these mathematical expressions apply only to single-stage units, a similar approach is used for multi-stage ones. Finally, as with other utility equipment, turbine efficiency (e.g. isentropic or thermal) can be calculated after optimisation.

2.3. Gas Turbines

Gas turbines are complex engines that consume different types of fuel and must compress ambient air in order to produce shaft power by expanding the combustion gases to (near) atmospheric pressure. These devices can be found in almost any new energy project driving electric generators and in recent years they have also become a very competitive option to run other rotational equipment. Since the performance of these units also depends on size and load, they are also modelled by combining the expressions for design and part-load operation:

$$Q_{f\,D}^{gt} = A^{gt} \cdot W_D^{gt} + B^{gt} \quad (13)$$

$$W^{gt} = C^{gt} \cdot Q_f^{gt} - W_{int}^{gt} \quad (14)$$

$$W^{gt} = C^{gt} \cdot Q_f^{gt} + (A^{gt} \cdot C^{gt} - 1)W_D^{gt} + B^{gt} \cdot C^{gt} \quad (15)$$

Where:

$Q_{f\,D}^{gt}$ Design (net) fuel input to the gas turbine (kW)

W_D^{gt} Design gas turbine shaft output (kW)

Q_f^{gt} Actual (net) fuel input to the gas turbine (kW)

W^{gt} Actual gas turbine shaft output (kW)

W_{int}^{gt} Willans line ordinate intercept (kW)

C^{gt} Willans line slope regressed from part-load operation (adim)

A^{gt} , B^{gt} Slope and intercept regression parameters for design operation (adim, kW)

Equation (15) represents a linear function to calculate actual turbine output from the amount of fuel being consumed and the size of the equipment (i.e. design output). It is important to note that the regression coefficients will vary for different types of gas turbines (e.g. heavyduty or aeroderivative) or fuels. Moreover, the conditions of the exhausted gases (mass flow and temperature) must be also determined in order to compute the potential of further energy use (e.g. steam production inside a HRSG):

$$M_{exh\,D}^{gt} = \alpha^{gt} Q_{f\,D}^{gt} \quad (16)$$

$$M_{exh}^{gt} = M_{exh\,D}^{gt} - \beta^{gt}\left(Q_{f\,D}^{gt} - Q_f^{gt}\right) = \alpha^{gt} \cdot Q_{f\,D}^{gt} - \beta^{gt}\left(Q_{f\,D}^{gt} - Q_f^{gt}\right) \quad (17)$$

$$T_{exh}^{gt} = T_{exh\,D}^{gt}\left(1 - k^{gt}\frac{Q_{f\,D}^{gt} - Q_f^{gt}}{Q_{f\,D}^{gt}}\right) \quad (18)$$

Where:

$M_{exh\,D}^{gt}$, M_{exh}^{gt} Design and actual GT exhaust flowrate (kg/s)

α^{gt} Regression parameter for design exhaust flow (kg/kJ)

β^{gt} Regression parameter for part-load exhaust flow (kg/kJ)

$T_{exh\,D}^{gt}$, T_{exh}^{gt} Design and actual GT exhaust temperature (°C)

Even though gas turbines can be operated in different ways, most units follow similar trends depending on equipment type and, hence, the regression parameters can be modified accordingly. Note that equation (18) is a non-linear expression and it is not included in the optimisation since the steam produced inside HRSG is calculated with another model. Finally, gas turbine performance will also vary considerably with ambient conditions and correction factors have been implemented in all the equations although they have been omitted to simplify the equations.

2.4. Heat Recovery Steam Generators

HRSG can be found in almost any large industry site either as direct component of a processing plant or attached to gas turbines in the utility system. These units employ the heat contained in a flow of waste gases to raise the thermal energy of water or steam and, although they perform a similar role as boilers, they have distinctive characteristics that require a separate examination. In order to calculate the steam output from a HRSG it is necessary to carry out not only an energy balance, but also check for temperature feasibility of the process. Thus, the actual steam generation from a HRSG attached to a gas turbine (GT) will have a complex behaviour at part-load as both exhaust temperature and mass flow from the engine will also vary. However, since steam production inside unfired equipment (i.e. GT + UF-HRSG) is limited by the minimum temperature approach between gas and steam/water streams (i.e. pinch point), it is possible to deduce a straightforward expression to determine the maximum mass flow that can be raised by these units:

$$M^{hr} = \frac{Cp_{exh}^{hr} M_{exh}^{gt}}{(\Delta h_{sh}^{hr} + \Delta h_{eva}^{hr})}(T_{exh}^{gt} - T_{sat}^{hr} - \Delta T_m^{hr}) \quad (19)$$

Where:

M^{hr} Net steam production in the UF-HRSG (kg/s)

Cp_{exh}^{hr} Average specific heat for the GT exhaust gases (kJ/kg-°C)

Δh_{sh}^{hr} , Steam enthalpy difference across the UF-HRSG superheater (kJ/kg)

, Δh_{eva}^{hr} Steam enthalpy difference across the UF-HRSG evaporator (kJ/kg)

ΔT_m^{hr} Minimum temperature difference between gas and steam streams in the UF-HRSG (°C)

Although equation (19) is a non-linear relationship, when it is combined with the expressions from the gas turbine model a linear function can be derived to establish steam production inside UF-HRSG at any load taking into account changes in GT exhaust flow and temperature:

$$M^{hr} = \frac{Cp_{exh}^{hr}}{(\Delta h_{sh}^{hr} + \Delta h_{eva}^{hr})} \begin{bmatrix} \left(T_{exh\,D}^{gt} - T_{sat}^{hr} - \Delta T_{m}^{hr}\right)\left((\alpha^{gt} \cdot Q_{f\,D}^{gt}) - \beta^{gt}(Q_{f\,D}^{gt} - Q_{f}^{gt})\right) \\ -k^{gt} \cdot T_{exh\,D}^{gt}\left(\alpha^{gt} - \beta^{gt}\right)\left(Q_{f\,D}^{gt} - Q_{f}^{gt}\right) \end{bmatrix} \tag{20}$$

Note that in equation (20) a negligible quadratic term has been removed to yield a perfectly linear relation. On the other hand, when supplementary firing (SF) is employed to increase HRSG output, it is possible to calculate maximum steam generation and the amount of fuel required with the following equations:

$$M^{sf} = \frac{Cp_{g}^{sf}}{\Delta h_{tot}^{sf}}(M_{exh}^{gt} + M_{f}^{sf})(T_{g1}^{sf} - T_{stk\,m}^{sf}) \tag{21}$$

$$M_{f}^{sf} = \frac{Cp_{exh}^{sf} \cdot M_{exh}^{gt}}{LHV^{sf}}(T_{g1}^{sf} - T_{exh}^{gt}) \tag{22}$$

Where:

M^{sf}	Net steam production in the SF-HRSG (kg/s)	M_{f}^{sf}	Required SF to reach T_{g1}^{sf} inside SF-HRSG (kg/s)
Cp_{g}^{sf}	Average specific heat for the hot gases after supplementary firing (kJ/kg-°C)	T_{g1}^{sf}	Maximum firing temperature in the SF-HRSG (°C)
Δh_{tot}^{sf}	Steam/water enthalpy difference across all SF-HRSG heat exchangers (kJ/kg)	$T_{stk\,m}^{sf}$	Minimum stack temperature for the SF-HRSG (°C)
		LHV^{sf}	Net calorific value of the supplementary fuel (kJ/kg)

Since maximum steam production (and maximum SF efficiency) can only be obtained by reaching high firing temperatures and minimum stack temperature will usually be also attained, equation (21) represents a linear function to determine the actual steam flow generated inside a SF-HRSG. Regarding the amount of supplementary fuel, again equation (22) is a non-linear expression that can be combined with the GT model to derive a linear function that takes into account flow and temperature variations in the exhaust gases:

$$M_{f}^{sf} = \frac{Cp_{exh}^{sf}}{LHV^{sf}} \begin{pmatrix} (T_{g1}^{sf} - T_{exh\,D}^{gt})(\alpha^{gt} \cdot Q_{f\,D}^{gt} - \beta^{gt}(Q_{f\,D}^{gt} - Q_{f}^{gt})) \\ +\alpha^{gt} \cdot k^{gt} \cdot T_{exh\,D}^{gt}(Q_{f\,D}^{gt} - Q_{f}^{gt}) \end{pmatrix} \tag{23}$$

3. PROPOSED METHODOLOGY

In order to handle varying conditions, a multiperiod optimisation has been implemented in which different scenarios can be defined. Thus, the resulting data determines not only a plant configuration that can cope with all the specified fluctuations, but also it establishes how to operate the system during each scenario. In this sense, the proposed approach adds a new dimension to the conventional design problem since the optimum solution will now depend on both structural and operating variables. As expected, the optimisation task becomes more complex but by employing the suggested models it is possible to generate a mixed-integer linear (MILP) formulation for which the objective function to minimise is total cost:

Objective function:

$$\min C_{Tot}(x, y, z_D, z_{op}) = \sum_{n} C_{cap}(x, z_D)_n + \sum_{t} C_{op}(x, y, z_{op})_t \tag{24}$$

Subject to:

$$h_j(x, y, z_D, z_{op}) = 0, \quad j = 1,2,\ldots,J \tag{25}$$

$$g_k(x, y, z_D, z_{op}) \geq 0, \quad k = 1,2,\ldots,K \tag{26}$$

$$x_n \in \{x^{\min} \leq x \leq x^{\max}\}_n \qquad z_D \in \{1, 0\}_n$$

$$y_{n,t} \in \{y^{\min} \leq y \leq y^{\max}\}_{n,t} \qquad z_{op} \in \{1, 0\}_{n,t}$$

Where:

C_{Tot}	Total utility plant cost (capital and operating)	C_{cap}	Operating cost of each n equipment unit in the plant along every t operating scenario
C_{cap}	Capital cost of each n unit of equipment in the plant		

x_n Maximum output (size) of each n unit in the plant

$y_{n,t}$ Actual output of each n unit in every t operating scenario

z_D Binary variable for the selection of each n unit in the plant

z_{op} Binary variable for the on/off status of each n unit in the plant along every t operating scenario

J, K Number of equality/inequality constraint functions

The major constraints for the objective function are energy and mass balances together with equipment performance equations. While the capital cost of each unit is a function of the selected sizes, operational costs will depend on equipment sizes and their actual loads. Performance of all units will translate into different quantities of overall fuel and water requirements, electricity import/export, and emissions for every operating scenario. Note that the emissions are obtained from stoichiometric relations upon the fuels being consumed. Whilst the mathematical formulation employs continuous variables to define, for example, equipment sizes and loads, binary variables are used to select the units of the plant and designate whether they operate or not during a certain scenario. Structural decisions are taken from a super-configuration that can include user specifications (i.e. flexible superstructure) to tackle retrofit or pure operational problems by fixing some or all the structural elements of the system.

3.1. Mechanical Driver Selection

Shaft power for rotational equipment (e.g. fans, compressors) might be provided by many different types of drivers and can considerably modify the final design and operation of the whole utility system because of the complex interactions with other equipment. Consequently, it is possible to achieve significant savings if large driving units are selected in an integrated context during the synthesis procedure. Nevertheless, conventional approaches to the design of utility plants only consider pre-specified types for drivers with fixed size. Thus, another innovative feature implemented in this work is an exhaustive driver selection to satisfy mechanical power demands.

Figures 2a and 2b illustrate how a large number of driver options can be included in the superstructure for configuration decisions. Shaft demands can be satisfied with gas turbines with or without HRSG, back-pressure or condensing turbines through different expansion paths, electric motors, or a combination of all of them. Moreover, the number and sizes of the units attached to a single shaft are also unknown. In this manner, the types, number, sizes and loads for all drivers are optimized together with the rest of the utility system.

3.2. Equipment Redundancy

The design of a utility system should also make allowances for equipment being unavailable due to maintenance and unexpected failure. As result, plants operate with some redundant equipment (i.e. spare or oversized units) most of the time. In conventional design routines a discretionary number of additional units or fixed oversizing factors are assumed to cope with these concerns. However, given the possibility of simultaneous structural and operational optimization, the proposed methodology can establish an optimal equipment redundancy and attain large savings in capital expenditure. For instance, users can define several scenarios and specify failure conditions in some of them, so that the final design will cope with all situations at minimum cost.

For instance, if the operating horizon of a plant is divided into maintenance time intervals and each unit is forced to be shutdown at least during one of these periods, then the optimizer will establish at the same time the operational planning (scheduling) and the redundancy due to maintenance that will minimize total cost. In addition, since two units failing independently at the same time is very unlikely, optimal redundancy due to unexpected failures can be also obtained by forcing in certain periods the shutdown of the two largest units of each type (i.e. one of them fails while the other one is receiving maintenance). Hence optimization might cope with such situations by adding new equipment, increasing the size of some units or by importing/exporting more/less utilities (e.g. electricity). It is worth mentioning that the size or the types of units that will be employed as redundant equipment is not known beforehand, so the procedure will determine the optimal equipment redundancy without the use of any heuristic or rule of thumb.

4. CASE STUDY

The capabilities of the proposed methodology are briefly demonstrated with a retrofit case. An existing utility system must satisfy a new set of demands due to the incorporation of another processing plant. Demands and site data are provided in table 1 for two seasons with different electric tariffs.

Table 1. Site and demands data for the study case.
Note that different electric tariffs apply to each season.

PARAMETER	UNITS	VALUE		NEW DEMANDS	
		Base	Summer	Base	Summer
Fraction of the year	%	67.0	33.0		
Ambient temperature	°C	10.0	25.0		
Electric prices: Peak	$/kWh	0.07	0.08		
Off-peak	$/kWh	0.05	0.05		
Peak hours per day	hrs	7.0	12.0		
Fuel oil cost	$/kg	0.19	0.19		
Natural gas cost	$/kg	0.22	0.22		
Raw water cost	$/ton	0.05	0.05		
Electricity demands	MW	62.0	68.0	81.0	89.0
Steam demands	MW	424	382	593	534
Condensate return	%	80.0	80.0	80.0	80.0
Power pump 1	MW	5.2	5.0	7.2	7.0
Power pump 2	MW	1.3	1.1	1.8	1.5
Power pump 3	MW	2.2	2.0	2.8	3.0
Power pump 4	MW	0.6	0.6	0.8	0.8
Process CW demands	MW	200	300	280	420

A conventional method to deal with the problem is to test many reasonable retrofit options with an operational optimizer and select the one yielding the best results. However, this approach can hardly evaluate all possible alternatives since equipment type, number of units, sizes and all their connections must be pre-specified by the user. A promising answer obtained in this way is presented in Figure 3a. New equipment is highlighted, two more boilers of the same size plus a back-up are provided and additional steam is expanded through new turbines to generate electricity or deliver mechanical power. Required investment is 62.0 MM$ and the operation cost is 111.0 MM$/yr. A different retrofit design results when the suggested methodology is applied as seen in Figure 3b. There are still three additional boilers but the third one is of a smaller size. Cogeneration potential from extra steam available is also exploited in turbines, but through different expansion paths. Required investment is only 39.0 MM$ with an operational cost of 108.8 MM$/yr yielding economic savings of 23.0 MM$ and 2.2 MM$/yr respectively. It is worth mentioning that several near-optimum answers can be generated in order to provide the user with several options to evaluate. For example, if more investment is allowed, it would be possible to include a gas turbine and the optimum answer would require 49.0 MM$, but would feature 5.8 MM$/yr of savings.

Figure 3: a) Promising retrofit solution for the study case obtained with conventional methods.
b) Optimal answer provided by the suggested methodology

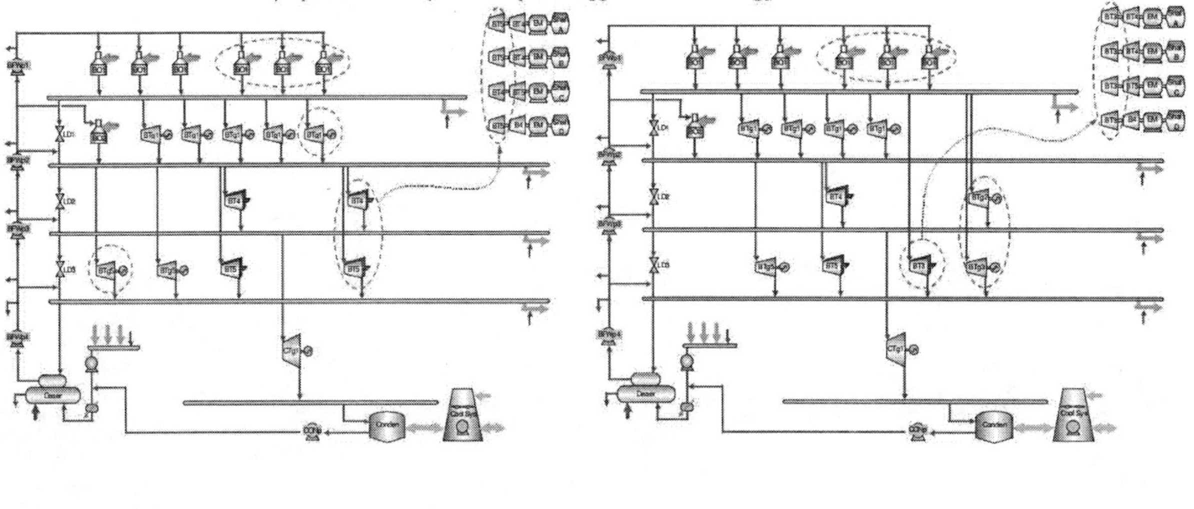

5. CONCLUSIONS

A novel methodology to optimize the design and operation of flexible utility systems has been developed. The proposed modeling framework allows simultaneous structural and operational optimization in order to tackle grassroots, retrofit and pure operating problems. Since the multiperiod MILP formulation can handle varying conditions in which equipment sizes and loads become variables, it is also possible to establish the optimal redundancy, maintenance planning for all units, and carry out an exhaustive driver selection that minimize total cost. In comparison with conventional methods, the suggested approach has the potential to achieve large economic benefits when applied to real industry problems.

REFERENCES

Bruno J. C., Fernandez F., Castells F. and Grossmann I. E. "A rigorous MINLP model for the optimal synthesis and operation of utility plants." Trans IChemE (Chem. Eng. Res. and Design). Vol. 76, Part A (1998): 246-258.

Hui C. and Natori Y. "An industrial application using mixed-integer programming technique: a multi-period utility system model." Computers and Chemical Eng. Vol. 20. Suppl. (1996): S1577-S1582.

Iyer R. R. and Grossmann I. E. "Optimal multiperiod operational planning for utility systems." Computerschem. Engng. 21.8 (1997): 787-800.

Maia L. O. A. and Qassim, R.Y. "Synthesis of utility systems with variable demands using simulated annealing." Computers and Chem. Eng. Vol. 21. No. 9 (1997): 947-950.

Maréchal F. and Kalitventzeff B. "Targeting the integration of multi-period utility systems for site scale process integration." Applied thermal engineering 23 (2003): 1763-1784.

Nath R., Holliday J. F. "Optimizing a process plant utility system." Mechanical Engineering February (1985): 44-50.

Papoulias S. A. and Grosssman I. E. "A structural optimization apprach in process synthesis - I. Utility Systems." Computers and Chem. Eng. Vol. 7. No. 6. (1983): 695-706.

Varbanov P. S., Doyle S. and Smith R. "Modelling and optimisation of utility systems." Chem Eng Research and Design 82.a5 (2004): 561-578.

Systematic driver and power plant selection for power-demanding industrial processes

Del Nogal, F.L., Kim, J., Perry, S.J. and Smith, R.

Centre for Process Integration
School of Chemical Engineering and Analytical Science
The University of Manchester
Sackville Street
Manchester M60 1QD, U.K.

Prepared for Presentation at the AIChE Spring National Meeting 2005
Atlanta, April 10-14, 2005
Energy Efficiency for a Sustainable Economy

Unpublished

ABSTRACT

This work presents a systematic methodology for driver and power plant selection based on mathematical programming. Optimal designs are generated given a set of mechanical and electric power demands, an economic scenario and other design constraints. The solution consists of the optimal number, type and actual size and model of the main compressor drivers, helper motors or generators and power plants, as well as the compressor stage arrangement. The discrete nature of gas turbines is considered and gas turbine drivers and pre-engineered power plants are selected from a group of candidates. The use of piecewise linearisation in some constraints maintains the model robustness, without leading to significant inaccuracies. Parallel compression sets are allowed in the formulation in order to benefit from the associated step improvement in plant availability. Plant availability has also been included in the objective function under the form of lost profit. Despite the associated non-linearities, this adds a new dimension to the classical approach and allows a more comprehensive exploitation of the trade-offs between capital costs, operating costs and availability.

When ignoring process heating and neglecting any potential steam system that is not part of a pre-engineered power plant, a simplified formulation can be applied to heavily power dominated processes, such as LNG. However, a more comprehensive formulation, allowing waste heat recovery and integration with a multilevel steam system, is able to produce more thermally efficient systems which become competitive under some economic scenarios despite their increased capital cost and complexity. Furthermore, addressing the supply of process heat via the steam system extends the applicability of the methodology to a wider range of process power-to-heat ratios.

The formulation has proved to be flexible and robust enough to produce solutions ranging from no-steam to all-steam systems. Its application to industrial case studies produces optimal and near-optimal systems for a variety of economic scenarios.

INTRODUCTION

A significant amount of work has already been published in relation to the systematic design of utility systems. Some approaches are based purely on thermodynamic analyses, focusing on maximising the energy efficiency of the system. Capital costs are not considered in these methods and therefore the resulting design is often not a true economical optimum, even when energy costs are dominant. This is the case in Nishio *et al.* (1980) and (1982), Chou and Shih (1987) and Zheng *et al.* (1996).

Mathematical programming allows the automated design of chemical processes at the conceptual level. Mixed integer programming has been particularly useful as it is able to handle discrete design decisions. These methods are based on a superstructure in which all the potential flowsheet alternatives are contained and from which the optimisation algorithm identifies the best combination. Papoulias and

Grossmann (1983) proposed an optimisation model for synthesis of utility systems based on mixed integer linear programming (MILP). Cost functions were linearised and operating conditions fixed to keep the constraints linear. Later, Bruno *et al.* (1998) formulated the same problem in a mixed integer non-linear programming fashion (MINLP), in an attempt to provide greater accuracy.

Maia *et al.* (1995) proposed an optimisation model based on a simulated annealing algorithm. They claimed to be the first to consider standard equipment sizes and costs. Wilkendorf *et al.* (1998) and Maréchal and Kalitventzeff (1998) also proposed methods capable of considering these discrete aspects. Manninen and Zhu (1999) proposed a multi-level decomposition of the problem. Continuous functions for gas turbine data were applied in a preliminary screening stage to identify the optimal size range. In a later stage, discrete gas turbine options were considered. Their methodology was, however, focused on retrofit of existing utility systems by integration of gas turbines and not on the design of new systems.

The problem has been also tackled with multi-period operation considerations, in order to take account of operational variations. Such is the case of Maia and Qassim (1997), Iyer and Grossmann (1998) and Shang *et al.* (1999).

The approaches above don't account for plant reliability in a systematic way. Also, they are based on steam superstructures and cannot afford solutions without the presence of steam. Gas turbines were included in some cases as an option for electricity generation and for heat recovery from their exhaust, but not for mechanical drive service. Steam turbines and electric motors were the only available options for satisfying mechanical demands and these were limited to drive only one piece of equipment and not a combination on the same shaft.

The provision of a steam system as part of the overall utility system is not always essential. Such is the case of power dominated processes (e.g. LNG), where the current trend is to use a hot oil circuit for process heating and a full steam system is considered an unnecessarily complex and expensive option. Nevertheless, running steam turbine drivers or generators with heat recovered from gas turbine exhaust could still be attractive when thermal efficiency and greenhouse emissions are of prime importance.

It is also essential to account for the discrete nature of some equipment (e.g. gas turbines and related standard power plants). This is reflected in the form of a set of available sizes, performance and costs. As Maia *et al.* (1995) point out, using continuous functions instead (e.g. from data regression) would lead to two possible situations once a solution is found. Firstly, if the equipment to be installed is intended to feature exactly the same characteristics as in the optimisation solution, then the high cost of special (custom) components might cause the design to be uneconomical. Secondly, if standard models with higher capacities are adopted, then the optimal conditions might be lost because the consequent higher costs have been neglected. Thus, it is critical that these discrete variables remain discrete in the formulation of the problem.

COMPRESSOR DRIVERS

Each compressor or compressor stage represents a demand for mechanical power, which has to be supplied by a driver, connected via a shaft. It is possible to run several compressors on the same shaft, hence consuming power from a common driver and running at the same speed. When the total mechanical power demand of a shaft is beyond a practical limit for the driver, it is necessary to use more than one shaft (more than one driver) in order to satisfy the compressor demands.

If several shafts are required, there will be different alternatives for assigning the compressors to the shafts and even for splitting stages belonging to the same compressor and running them on different shafts. Consecutive stages belonging to the same compressor that run on the same shaft could be built into a single casing provided they are of the same nature (e.g. centrifugal), which is an economic incentive. The lowest compressor cost would result from having one casing. However, the power supplied by each driver must be in balance with the sum of the demands on its respective shaft. This constraint, together with a practical upper limit for the driver size, means that sometimes a large compressor must be split into several casings to allow being driven by more than one shaft.

Steam Turbines

Steam turbines have been the classic choice for drivers in conventional utility systems for many years. In power dominated processes, the use of steam turbine drivers was a common practice in the past, providing all of the required power. The power system design was relatively straightforward as steam turbines can generally be manufactured to customer requirements.

The need for a complete steam system to support the operation of such turbines has been the main drawback of this option for power dominated processes, since water treatment and steam production, distribution and condensing represent a significant capital investment (Bellow *et al.*, 1997 and Liu *et al.*, 1995). Given that the heating demands in these processes are minor, gas turbine drivers replaced steam turbines as the driver of choice since they became common practice. Steam could then be removed from the plant and process heating provided via alternative systems such as hot oil circuits.

Under the current worldwide trend towards more environmentally benign designs, steam turbine drivers are likely to return as a standard practice in power dominated processes. Being supplied with steam recovered from the exhaust heat of gas turbine drivers, they can be useful to push the limits of thermal efficiency (Peterson *et al.*, 2001 and Avidan *et al.*, 2002).

Gas Turbines

Gas turbine systems are more efficient for standalone power generation, hence requiring less fuel and producing fewer emissions per unit of product than steam-based power systems. The acceptance of gas turbine drivers is also due to the increased reliability, operating range and operating experience of the industrial gas turbine as well as its relatively easy start up (Bellow *et al.*, 1997 and Aoki *et al.*, 1995). Aero-derivative gas turbines, more compact and efficient, are still unproven in these applications but, nevertheless, several recently announced LNG projects have plans to use them (Avidan *et al.*, 2002).

Unlike steam turbines, gas turbines are available only in discrete sizes. Although in principle custom designed gas turbines could be developed, the cost of doing so would certainly be impractical for the purchaser. Gas turbines are rather a package concept and each model corresponds to a predefined set of components defining its performance (air compressor, combustor, expander, lube oil system, etc.). The overall trend is that the larger the turbine, the higher its efficiency and the lower its specific cost. Aero-derivative gas turbines provide a significant improvement in efficiency but at a higher specific cost than industrial ones.

Performance data of gas turbines is normally available for ISO conditions. When atmospheric conditions or the fuel composition are different so will be the performance of the gas turbines. Power output and other performance parameters have to be adjusted in order to reflect the real operating conditions. GE Power Systems has published some plots where de-rating factors can be estimated (Brooks, 2000). Additional factors should also be considered to account for the effect of fouling and ageing. Coyle *et al.* (1998) estimate that the air compressor fouling in gas turbines occurs very quickly and that a constant factor of 2% can be applied. They also report a factor of 4% normally taken for ageing, which occurs rapidly during the first year. Also, bear in mind that the performance of gas turbines deteriorates significantly at partial load. Using gas turbines at partial load during normal operation is not a common practice because of their unattractive efficiencies and the inappropriate capital utilisation. Usually no gas turbines larger than frame 7 (~80 MW) are considered for mechanical drive applications because it would likely lead to a significant decrease in plant availability.

Electric Motors

The use of electric motor drivers will result in a simpler application as they do not need as many support systems as gas turbines do (air compression, combustion, fuel gas, etc.). If the plant is not able to import electricity, it will be necessary to have a power plant on the site and is evident that some of the saved complexity will be transferred to the gas turbines used there. It should be considered, however, that there will always be some electricity demand, in which case the gas turbine support systems would be kept to just one centralised application, which is the power plant (Richardson *et al.*, 2001).

If the electricity were to be generated on site, the use of electric motors would lead to a higher overall thermal efficiency since very large gas turbines (e.g. frame 9) could be used for electric generation. These have lower specific fuel consumption and capital cost than the smaller direct drive gas turbines (DDGTs). This could lead to a decrease in atmospheric emissions per unit of power generated and perhaps in the total capital investment, although the extra cost for the motors themselves as well as for the electric distribution (transformers, transmission lines, etc.) must be considered. A combined cycle power plant can achieve even further improvements in the overall thermal efficiency and greenhouse emissions.

The practical power limit for a single 4-pole synchronous motor is around 60 MW and the largest 2-pole synchronous motor built, until recently, is around 53 MW (Fallaize and Phillips, 2002). Therefore, the use of electric motors around these sizes and beyond involves certain technological risk. Precedents exist in the air separation industry employing motors up to 52 MW. The Snøhvit LNG project, in Norway, is a pioneering move of the LNG industry towards large electric motors. 65 MW electric motors fed by several aero-derivative gas turbine generators will be used and a connection to an external grid will also be implemented (Shu *et al.*, 2002).

Helper Motors and Generators

Gas turbines themselves require a smaller driver to provide the torque needed during the start up of the compression system. A variable speed electric motor is normally used as the starter driver. Starter motor sizing is not a trivial task. A dynamic simulation of the system during start up might be required to do so (Heckel and Davis, 1999).

It is often economically attractive to use this installed capability (starter motor) in a continuous fashion as a helper to the gas turbine, supplementing the turbine power output and maximising the production capacity with the installed equipment (Liu *et al.*, 2001 and DiNapoli and Yost, 1998). Starter motors could also act as electric generators in case there is excess power in the shaft (Vink and Nagelvoort, 1998 and Kleiner *et al.*, 2003).

The use of continuous size elements, such as helper motors or generators, makes it possible to keep the shaft in balance and to smooth the effect of the discrete nature of gas turbines on the design of the power system.

Electricity Generation

In addition to the mechanical demands, there will also be some electricity required to run smaller equipment such as fuel and boil-off gas compressors, fans, pumps, instrumentation, utilities, etc. If the plant is not able to import electricity, an on-site generation system will be required.

Apart from a classic pure-steam system, two other options exist for an on-site power plant. The first option is a simple cycle power plant, which is in essence a gas turbine running an electric generator with no use of the exhaust gas heat. The second option is a combined cycle power plant. These plants consist of a gas turbine (or several) running an electric generator(s) as in the simple cycle case, but in addition steam is raised with the exhaust gas heat in order to increase the generation with the use of steam turbines. The thermal efficiency is, of course, improved by recovering heat from the exhaust gas. Small to very large gas turbines, such as frame 9's, are suitable for electric generation in both types of plant.

One could claim that combined cycle plants imply a steam system and thus high capital costs as previously mentioned. However, like gas turbines, these power plants are basically discrete options: a pre-engineered and self-contained package, which helps in keeping the steam system costs under control.

Again, there is a trend for the efficiency and specific cost of simple and combined cycle power plants to improve with size. Combined cycle options offer a step improvement in efficiency but have higher specific capital costs than simple cycle plants.

Plant Capacity and Power Balance

If the power system is based purely on steam, it could be simply custom made according to the power process requirements. Once the plant capacity is chosen, the characteristics of the driven equipment are obtained and a set of specifications for the steam turbine drivers can be produced. The same procedure applies for the steam turbines intended for electricity generation. The continuous size nature of steam turbines allows this quite straightforward procedure.

The situation is different when gas turbines are part of the power system. The use of helper motors or generators could eventually help in balancing the power in individual gas turbine shafts. Also, when using main electric motor drivers, there should be no major difficulty in balancing the shafts as well. However, if gas turbines are involved in the electric generation, either in simple or in combined cycle, the overall energy balance will rely on discrete sized elements. It will be more difficult then to achieve an exact global match between the power demands and sources.

Importing or exporting electricity could be an option to achieve overall power balance but this is not always possible either because of a remote location (e.g. offshore) or because it is not economically attractive. In this type of situation, and for a given production rate, the designers would often have to select a power system with excess capacity. One could choose to run the power system at partial load in order to achieve the proposed plant capacity, but it would imply poorer gas turbine performance besides an incomplete utilisation of the invested capital. To avoid this, the plant capacity is sometimes adjusted in power dominated processes in order to match the available power (Coyle *et al.*, 1998, De La Vega *et al.*, 1995). The overall plant capital investment will increase since the power system remains the same, but the rest of the plant has to be designed for higher flowrates. Nevertheless, capital utilisation and specific capital expenditure, fuel consumption and greenhouse emissions would be more favourable.

PLANT AVAILABILITY

Plant availability has a major impact on process economics, especially when a significant number of rotating equipment is involved. Unscheduled downtime is associated with the successful performance of heat exchangers, pumps, valves, instrumentation and control, utilities, human factors, etc. The choice of rotating equipment and compression strategy can have a significant influence on the amount of time that the plant is down or operating at partial load. Designers must be aware of the availability implications in the choice of rotating equipment and their arrangement. Initially, one power system design could appear to be more appropriate than others based on equal operating time, but if it implies a lower availability, it might lose such an advantage.

Reliability figures deal with unplanned events and generally exclude all outages associated with scheduled maintenance activities. Usually the larger the piece of equipment, the lower its reliability. Scheduled downtime is another factor that affects the annual plant production. Again, usually the larger the piece of equipment, the longer its scheduled maintenance time. Overall, large items of equipment are able to bring the benefit of economy of scale, but one must be aware of the consequent increased scheduled and unscheduled downtime that they imply (decreased availability).

Equipment redundancy has a positive impact on plant availability and is employed usually to overcome any relatively unreliable elements in the system (e.g. a spare pump or parallel valve). An N+1 sparing philosophy is frequently used for boilers and power generation units. That is, if N units are required to run at full capacity to supply the normal demand, then invest capital in the installation of one extra unit able to cover the deficit left by the failure of any one of the others. The spare unit may either be kept offline to be used only in case of contingency or online, running simultaneously with the other N units at partial capacity. In both cases, some capital utilisation is sacrificed to guarantee a reliable operation.

Redundancy may also take the form of parallel compression, as illustrated in Figure 1. That is, splitting each low-pressure stream in two equal parts, before being sent to compression, and feeding two independent driver-compressor sets. This strategy would allow a process to benefit from the economy of scale while still having good availability. In case one compressor or driver is down, the plant could keep producing at partial load.

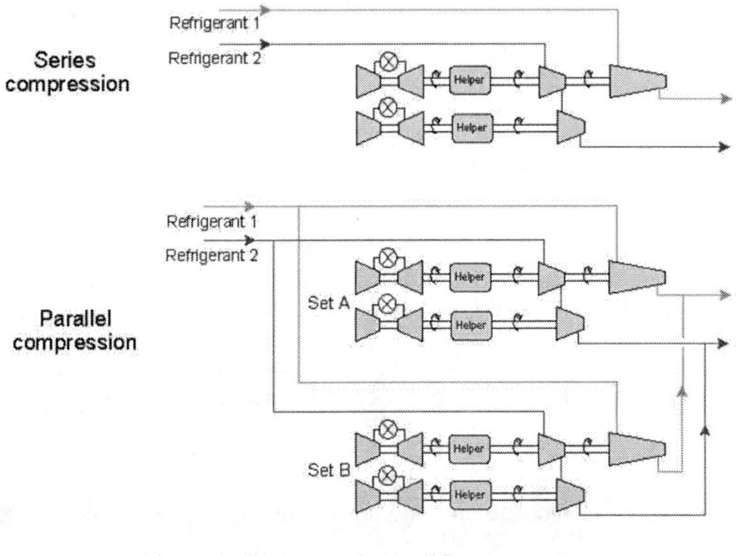

Figure 1. Series and parallel compression.

A NEW APPROACH

Designing a power system involves discrete decisions, not only because of the discrete options available for gas turbines and power plants, but also because of the discrete power demands and arrangement alternatives of the components in the system. Maximum utilisation of the installed capacity is a key issue as well. The start-up motors of gas turbines are now often used as helper drivers to boost shaft power. The optimal power system must satisfy a set of mechanical, electrical and heating demands by employing the most cost-efficient number, type and model of mechanical drivers, compressor stage arrangement, helper motors or generators and power plants, either in simple or combined cycle. A proper optimisation model for this problem must allow for the flexible arrangement alternatives of driven equipment on different mechanical shafts in order to consider the different compressor casing alternatives. It must also be able to consider direct drive gas turbines and the use of helper motors or generators as balancing elements.

Superstructures

Figure 2 shows, in two parts, the superstructure on which the optimisation model is based. Figure 2a shows all the elements not requiring a customised steam system (steam may still be present in combined cycle power plants). That is, gas turbine drivers, electric motors, mechanical demands, power plants and the relevant electrical connections. Helper motors or generators are allowed on each gas turbine shaft. Each compressor stage is allowed to run on any of the gas turbines or motor shafts, hence allowing each compressor to be split into different casings. Figure 2b shows the steam-based elements in a complementary superstructure and how they are linked to the non-steam superstructure, in Figure 2a, through potential heat recovery steam generators at each gas turbine driver and simple cycle power plant. This multilevel steam superstructure includes steam generators, such as boilers and supplementary fired HRSGs, which may produce steam at any one of the existing levels, extraction steam turbine drivers and generators, letdown valves, vents, deareator and connections to provide/receive steam to/from the process at each level. Heat may be provided to the process through steam or through an implicit hot oil system. This makes the overall superstructure flexible enough to represent systems from no steam-at-all to pure-steam ones.

There will be a maximum number of potential gas turbines, steam turbines and main motor shafts that the designer will define in order to avoid excessive use of binary variables when solving the problem. Later, the solver will yield the most cost-effective number and type of these items. The designer can also define a maximum number of potential power plants.

The goal of the optimisation is to find a feasible power system with minimum total cost. This total cost may comprise any combination of capital costs, fuel costs, CO_2 emission penalties, lost profit due to unavailability, boiler water treatment operating costs and charges for the electricity imported. The latter would actually become a credit if electricity were exported instead. Taking into account all these costs will allow the model to trade-off the capital investment against the expected performance of the system. For the sake of a fair trade-off, the capital cost must represent the installed cost of the equipment and not just for the FOB cost. In this way the real investment is reflected and capital and operating costs are weighted more properly in the objective function.

A detailed description of the mathematical programming model is omitted from this paper as it would increase its length significantly. An early version of the problem formulation has been published in detail before (Del Nogal *et al.*, 2003), although important changes have taken place since then.

Availability issues are dealt with in the problem formulation in three different ways. First, an N+1 sparing philosophy applicable to the power generation units (power plants and steam turbine generators) can be requested to the model when defining the problem, just by activating a flag parameter before the optimisation. Secondly, the contribution of drivers and compressors to plant unavailability can be estimated in the model and translated to lost profit, which would represent an additional cost in the objective function. Lastly, with few modifications, an equivalent formulation can be also used to find optimal power systems that feature parallel compression.

In order to obtain the optimal solution, the total cost defined by the objective function must be minimised subject to all the relevant constraints (general formulation), which involve several non-linearities. The problem, in principle, is then of the Mixed Integer Non-Linear type (MINLP). A valid and consistent sub-problem can also be defined if the constraints dealing with the prediction of availability are not included (100% availability sub-formulation). Another way of neglecting plant availability would be to use the general formulation but with the individual reliabilities of all compressors, drivers and power plants set to 1 and their maintenance time set to 0 days/yr.

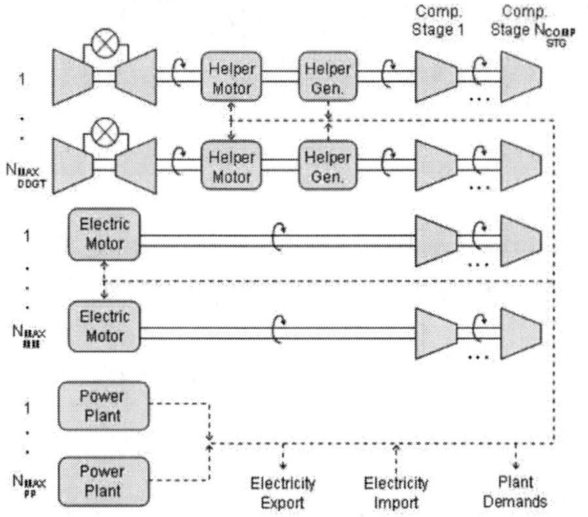

a) No-steam superstructure

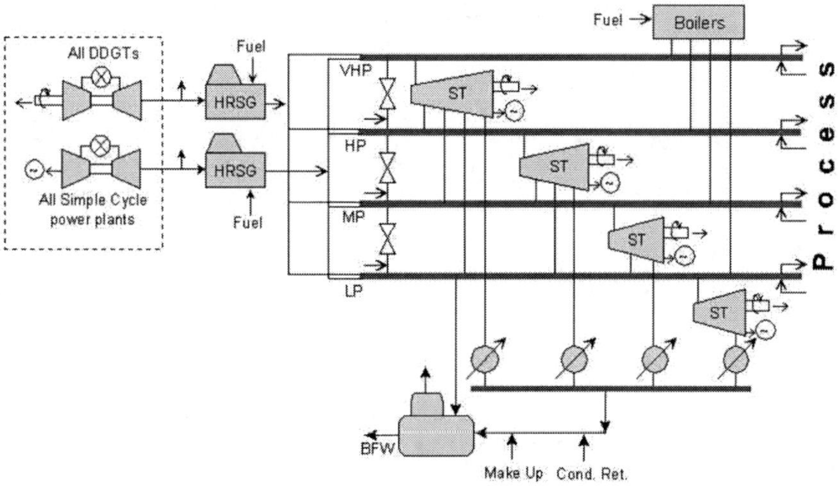

b) Steam superstructure.

Figure 2. Superstructures for the synthesis of the power system.

Linearising the problem

Non-linearities in the general formulation come from some of the equations dealing with cost estimation, equipment modelling and plant availability. Linearising the latter could result a very lengthy and complex task, if possible at all, but certainly the others can be linearised with a fair amount of work and extra equations. In doing so, the 100% availability sub-formulation could be transformed into an equivalent MILP problem, thereby being much easier to solve. Moreover, global optima would be guaranteed. Also, the solver robustness would be significantly enhanced when solving the general formulation, even when availability-related non-linearities force the problem to remain MINLP.

Most cost equations contain factors of the type X^R, where X is a variable and R a constant. Replacing each of these functions by a single straight line would result in a significant loss of accuracy. Instead, a more precise approximation, yet linear, can be implemented by combining several straight lines, each one valid over a given interval. That is, a piecewise linearisation (see Figure 3). Depending on how wide the expected range of equipment sizes is, it could be appropriate to use a larger or a smaller number of linear segments. The more segments, the more precise the approximation. Nevertheless, it must be kept in mind that these cost models are an approximation themselves and that using an excessive number of segments will just bring extra complexity with no significant improvement in the accuracy of the calculations.

There are various ways of performing piecewise linearisation. Perhaps the most compact formulation will be obtained by employing SOS2 variables (special ordered sets of type 2), which are supported by GAMS, the development environment used in this work (GAMS Development Corporation). Within an SOS2 set of variables at most two can have positive, non-zero values, and these two variables must be adjacent in the ordering given to the set.

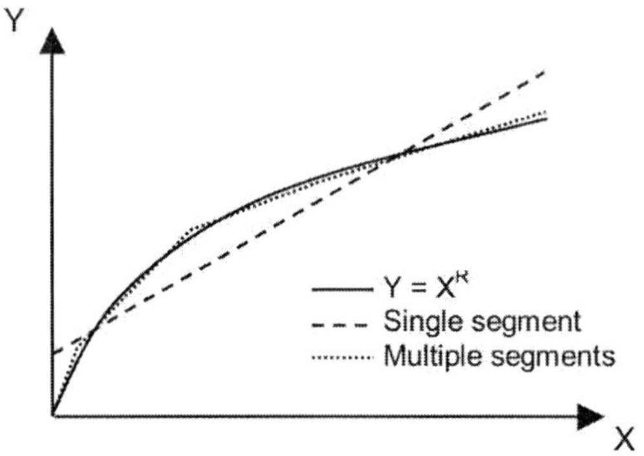

Figure 3. Linear approximation of a function of the type $Y = X^R$ (e.g. $0 < R < 1$).

With the purpose of illustrating the application of SOS2 variables to piecewise linearisation, let Y = f(X) be the function to be approximated and let the set of N data points (X_i, Y_i) define the N-1 linear segments for the approximation. That is, each segment being defined by each pair of consecutive (X_i, Y_i) points. Let also $\lambda = (\lambda_1 ... \lambda_N)$ be the SOS2 variable set. The linearisation is then performed by substituting Y = f(X) by the following three equations:

$$Y = \sum_i^N \lambda_i \cdot Y_i$$

$$X = \sum_i^N \lambda_i \cdot X_i$$

$$\sum_i^N \lambda_i = 1$$

Another important source of non-linearities to overcome is the modelling of steam turbines. Figure 4 illustrates a generic extraction steam turbine and the basic relevant nomenclature. An energy balance around any of the steam turbine stages or around any of the steam mains would feature the product of flowrates and enthalpies. In order to keep the formulation linear, the enthalpies of the steam mains are made fixed parameters ($\overline{hLv}_{Lv}$) by assuming constant pre-defined conditions, since they could be calculated before optimisation. As a consequence of this assumption, solutions may feature small amounts of steam letdown and/or boiler feed water injection to the steam mains in order to absorb any deviation from the chosen conditions. The outlet enthalpies from each steam turbine stage would also be variables in principle as, being rigorous, the performance of steam turbines (e.g. isentropic efficiency) is a function of both the expansion zone and the turbine size (Varbanov *et al.,* 2004). However, by neglecting the effect of turbine size, and as the steam levels are known before optimisation, one can estimate the steam turbine stage efficiencies and the stage outlet enthalpies by using, for instance, the model by Varbanov *et al.* (2004) also before optimisation (e.g. assuming a moderate steam flowrate). This permits keeping the formulation linear, since the stage outlet enthalpies, now parameters ($\overline{hExp}_{Lv,LvOUT}$), would depend only on the header conditions, which are already fixed.

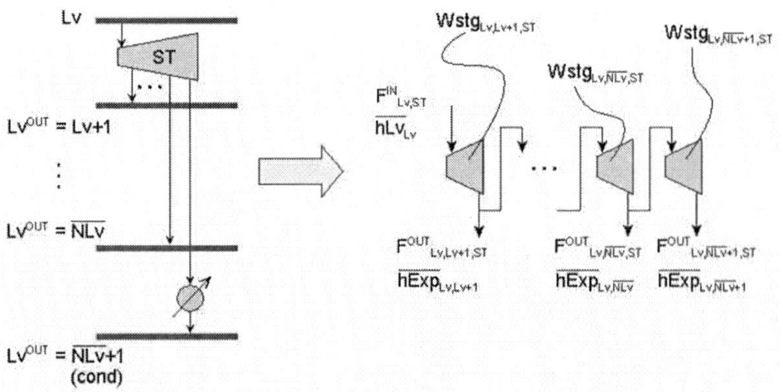

Figure 4. A generic extraction steam turbine.

CASE STUDY

The MILP version of the optimisation model was implemented using GAMS. All the process data and input parameters were provided in an interface spreadsheet and transferred automatically to GAMS using a Visual Basic macro in order to centralise the information and to make the scenario management easier.

Data

This case study is based on the APCI's Propane Pre-cooled Mixed Refrigerant LNG process. The process features a four level propane pre-cooling cycle, a mixed refrigerant cycle and an end flash system for nitrogen rejection. The rejected stream will provide most of the fuel required (end flash fuel).

A basic flowsheet and the set of process data for this case study were derived from a publication by Vink and Nagelvoort (1998). A simplified process flowsheet is shown in Figure 5. The mixed refrigerant compressor has three stages. The first one is axial and the rest centrifugal. It features three air fan coolers. The propane compressor has four centrifugal stages with side loads and three air fan coolers. In order to work from a consistent and relevant set of power demand data, the process as published by Vink and Nagelvoort was simulated. The basis was a single liquefaction train producing 5 mtpa (millions of metric tons per year) of LNG, which reflects current economics. The resulting refrigerant compressor demands are given in Table 1. The basic electricity demand is estimated as 42.7 MW, which does not include any allowance for potential electric motors or steam support systems. The process heating demands are estimated to be 34.2 t/h of medium pressure steam and 26.3 t/h of low pressure steam. That is, 36.31 MW of heat in total.

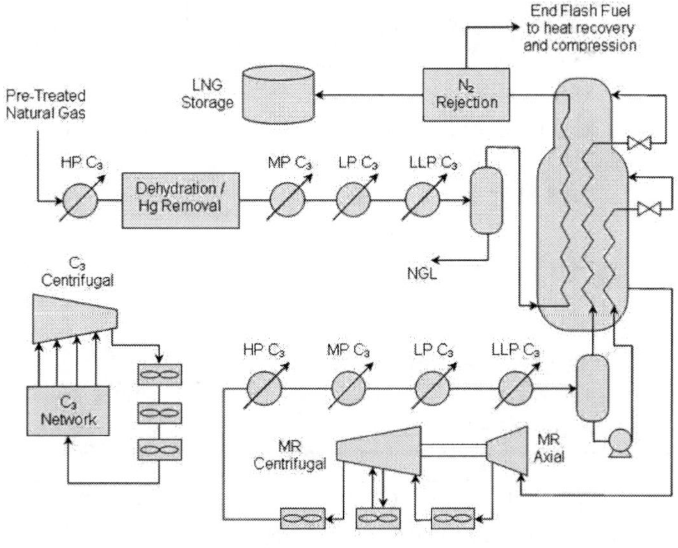

Figure 5. Simplified flowsheet of the APCI's Propane Pre-cooled Mixed Refrigerant Process.

Table 1. Mechanical power demands of the refrigerant compressors.

Compressor/Stage	Power Demand (MW)
C3/1	2.50
C3/2	6.96
C3/3	12.35
C3/4	32.25
Total C3	54.06
MR/1	57.64
MR/2	18.74
MR/3	27.34
Total MR	103.72

Other parameters of the base case scenario are summarised in Table 2. Candidate DDGTs and power plant choices in a pre-qualified list were de-rated to account for ambient temperature (25 °C), ageing, fouling and fuel composition. The fuel gas is initially assumed as a blend of 80% end flash gas and 20% pre-treated natural gas. This has to be checked in the solution for consistency. If significantly different, an updated fuel de-rating factor must be applied.

Table 2. Values of some parameters for the base case scenario.

Parameter		Value	
Equipment	Maximum main motor driver	60	MW
	Maximum steam turbine driver	60	MW
	Maximum helper motor size	25	% of GT
	Maximum helper generator size	25	% of GT
	Power required for DDGT start up	15	% of GT
	Motor efficiency	95	%
	Helper generator efficiency	95	%
	Mechanical transmission losses	1.5	%
Steam	Boiler and HRSG blowdown ratio	3.0	%
	Steam losses	2.0	%
	VHP Steam Level	490 / 85	^{o}C / bar
	HP Steam Level	390 /40	^{o}C / bar
	MP Steam Level	280 / 15	^{o}C / bar
	LP Steam Level	140 / 3	^{o}C / bar
	Condensing mains pressure	0.20	bar
	Minimum temperature difference	20	^{o}C
	Minimum stack temperature	130	^{o}C
Electricity	Electricity distribution losses	2	%
	Maximum electricity export	25	MW
	Maximum electricity import	0	MW
	Price of sold electricity	0	$/kWh
	Cost of imported electricity	0	$/kWh
Fuel and CO_2	Specific CO_2 emissions	0.199	kg/kW·h (LHV)
	CO_2 emissions penalty	0	$/tonne
	Cost of end flash fuel	0	$/kW·h (LHV)
	Cost of fresh fuel	1	$/kW·h (LHV)
	Max. end flash fuel available	540.3	MW (LHV)
Other	Plant lifetime	25	yr
	Interest rate	4	%

The plant will not be allowed to trade electricity and consequently its import is set to zero. There is no benefit from exporting electricity but the electricity export has not been banned in order to let the overall electricity balance close. The plant is intended to operate for 25 years. 350 operation days per year are assumed initially. Regarding the cost of compressors, a 20% penalty is assumed per additional casing.

Base Case Solution

The base case corresponds to minimising the total cost of the system considering a fresh fuel cost of 1 $/MMBtu and the end flash fuel at no cost. Only gas turbines and electric motors are considered as drivers in the first part of this case study. In that case, heat is supplied to the process through an implicit hot oil circuit. Figure 6a shows the optimal power system under the conditions of the base case scenario. A frame 7 and two aero-derivative gas turbines drive the compressor stages while a combined cycle is the selected power plant. The total cost of this system is 211.00 MM$ (millions of US$), out of which 206.81 MM$ are for capital expenditure and the remaining corresponds to the present value of the fuel costs along the plant lifetime. Taking the cost of fresh fuel as 1 $/MMBtu has encouraged the solver to select two aero-derivative gas turbine drivers and a combined cycle power plant to improve the thermal efficiency of the system.

The solution in Figure 6a features three dissimilar types of gas turbine drivers. This is often undesirable since it would imply different maintenance regimes and spare parts. Without great difficulty, extra constraints can be added to the formulation to limit the gas turbine driver types to a predefined maximum. The same can be done regarding power plants. Figure 6b shows the optimal power system for the base case scenario subject to identical gas turbine drivers and power plants. The solution features two identical frame 7 gas turbine drivers and a combined cycle power plant. The total cost is now 225.98 MM$ (7.1% higher), out of which 193.38 MM$ (6.5% lower) are capital costs. The absence of aero-derivative gas turbines allows a lower capital cost. However, for the same reason, fuel costs are now significantly higher and definitely overcome the capital savings. Aero-derivative gas turbine drivers cannot participate in this solution as identical drivers are being enforced and they are not large enough to drive the first stage of the mixed refrigerant compressor. The remainder of this case study will assume no constraints on the number of dissimilar equipment.

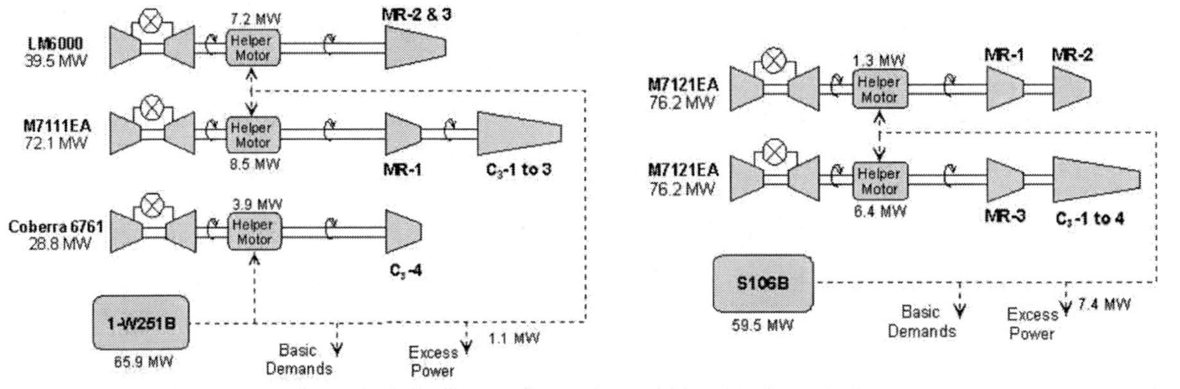

a) Unconstrained number of dissimilar equipment. b) Identical gas turbine drivers and power plants

Figure 6. Optimal no-steam solutions for the base case.

Sensitivity to fuel costs

A sensitivity analysis was performed considering five different fuel cost scenarios. The detailed results are reported in Table 3. As it is to be expected, the higher the fuel penalties, the more efficient the system becomes. The reduction in the total fuel consumption is achieved with the progressive use of aero-derivative gas turbines and combined cycle power plants. As discussed before, motor drivers increase significantly the electricity demand, which opens the opportunity of using larger and more efficient power plants. Such is the case in scenarios 3, 4 and 5. The capital costs are higher for the more efficient systems, as it is also expected.

Table 3. Sensitivity of the optimal power system to fuel costs.

Scenario	**1**	**2 (base case)**	**3**	**4**	**5**
Fresh fuel ($/MMBtu)	-	1.0	2.0	3.0	5.0
End flash fuel ($/MMBtu)	-	-	1.0	2.0	3.0
Driver Arrangement *	M7111EA+13.4MW HG for MR-1 M7121EA+2.3MW HG for C_3-1 to 4 & MR-2 M6581B+6.0MW HG for MR-3	Coberra6761+3.9MW HM for C_3-4 M7111EA+8.5MW HM for MR-1 & C_3-1 to 3 LM6000+7.2MW HM for MR-2 & 3	54.88MW Motor for C_3-1 to 4 58.5 MW Motor for MR-1 LM6000+7.2MW HM for MR-2 & 3	54.88MW Motor for C_3-1 to 4 58.5 MW Motor for MR-1 LM6000+7.2MW HM for MR-2 & 3	58.52MW Motor for MR-1 46.78MW Motor for MR-2 & 3 54.88MW Motor for C_3-1 to 4
Power generation	1xPG5371PA	1xPG5371PA	1xS109E	1xS109E	1xGUD1S84-3A
Total fuel (MW)	669.7	549.65	451.70	451.70	441.02
Capital cost (MM$)	186.19	206.81	228.10	228.10	242.05
Fuel cost (MM$)	-	4.19	202.13	404.25	592.04
Total cost (MM$)	**186.19**	**211.00**	**430.23**	**632.35**	**834.09**

* HM = helper motor, HG = helper generator.

Availability considerations

When obtaining the optimal parallel compression solution for the base case, and as the compressor demands have been halved in each set, the solver would be able to accommodate all the mechanical demands using aero-derivative gas turbine drivers this time. The total cost of the resulting parallel system (capital + fuel) is 217.18 MM$, which is slightly higher than the base case solution in series. However, once the lost profit due to the unavailability of drivers and compressors is taken into account, the parallel compression solution improves the series solution by 9.1% (assuming a profit of 50 US$ per metric ton of LNG). Furthermore, when availability is made part of the objective function as the cost of lost production, a much more dramatic result occurs. A new all-motor system in series is able to improve the base case solution in series by 45.0% and the base case solution in parallel by 39.5%. Details are shown in Table 4. Of course, an all-motor system would translate virtually all the availability issues to the electricity generation. The optimisation model was then run again considering N+1 sparing philosophy. Once the flag for power plant sparing philosophy is activated, the resulting series system with minimum total cost (capital + fuel + lost profit) is still an all-motor one, but this time an LM6000PC and a GUD2-64-3A (plus one spare) units are the selected power plants. Although the total cost is now 400.05 MM$ due to the capital increase, the availability improvement is still significantly dominant.

Table 4. System comparison considering lost production.

System	**Base case in Series**	**Base case in Parallel**	**Extended objective function in Series**
Driver Arrangement *	Coberra6761+3.9MW HM for C_3-4 M7111EA+8.5MW HM for MR-1 & C_3-1 to 3 LM6000+7.2MW HM for MR-2 & 3	Two identical sets of: Coberra6761+1.3MW HG for C_3-1 to 4 Coberra6761+0.5MW HM for MR-1 Coberra6761+5.3MW HM for MR-2 & 3	58.52MW Motor for MR-1 46.78MW Motor for MR-2 & 3 54.88MW Motor for C_3-1 to 4
Power generation	1xPG5371PA	1xLM6000PC	1xLM2000 1xS206FA
Capital cost (MM$)	206.81	213.52	246.94
Fuel cost (MM$)	4.19	3.66	-
Lost profit ** (MM$)	325.81	270.85	48.33
Total cost (MM$)	**536.81**	**488.03**	**295.27**

* HM = helper motor, HG = helper generator.

** Assuming profit of $50 per metric ton of LNG.

Steam systems

So far, steam systems have been neglected, as only the base superstructure in Figure 2a has been used. Extending such a superstructure to include all the steam-related options (in Figure 2b) results in a new optimal solution to the base case, which is shown in Figure 7. It has a total present cost of 195.5 MM$ (7.3% lower than the optimal no-steam solution), out of which 193.8 MM$ are for capital expenditure and the remaining for fuel and water treatment costs. However, the system is now more complex. It features one simple cycle power plant as the main source of electricity and a gas turbine frame 7 as driver for the third stage of the mixed refrigerant compressor and the whole of the propane compressor. Heat is recovered from both gas turbines in supplementary fired HRSGs in order to raise 373.5 t/h of VHP steam, which is enough for running the two steam turbine drivers and to provide the required process heating through MP and LP steam. The steam system features three steam levels (VHP, MP and LP) and a condensing main. Both steam turbines have an extraction stream to an intermediate steam level and the remaining steam is expanded down to the pressure in the condensers.

Steam vs. no-steam

For each of the fuel cost scenarios defined in the sensitivity analysis, the optimal general solution, the optimal no-steam solution and the optimal pure-steam solution were obtained. The results are shown in Figure 8 in the form of relative total costs. The first thing to notice from Figure 8 is how expensive the pure-steam optima result when compared to the no-steam optima and the general optima in all the five scenarios considered. They feature total costs from around 30% to more that 90% higher than the respective general optima. As was to be expected, the general optimal solution to Scenario 1 turned out to be a no-steam solution, since only capital costs are considered in the objective function. The details for Scenario 2 have already been discussed previously, where a hybrid gas turbine/steam turbine driver solution was the best option. A hybrid gas turbine/steam turbine driver solution still dominates in Scenario 3, but this time the advantage over the no-steam solution is smaller. In Scenario 4 the best choice is a no-steam solution, featuring gas turbine and motor drivers. Finally, in Scenario 5, the one with the highest fuel penalties, also a no-steam solution results as the best choice, this time with only motor drivers.

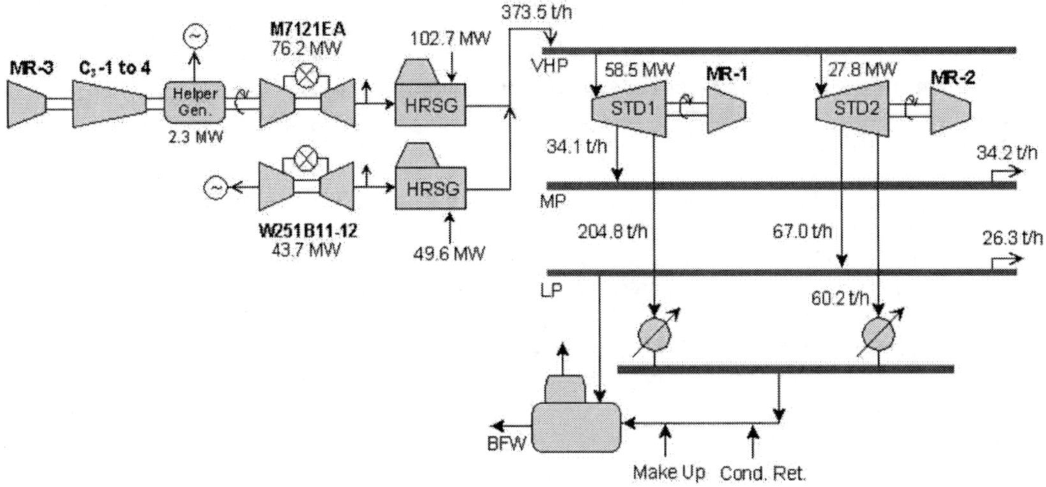

Figure 7. Optimal general base case solution.

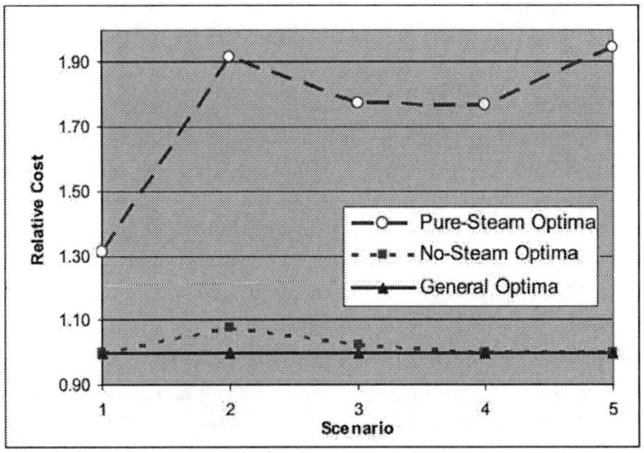

Figure 8. Steam vs. no-steam systems. Relative cost comparison.

In summary, the overall trend is that no-steam solutions are the best choice either at very low or at high fuel costs. In the first case they are dominant because of their relatively low capital cost. In the latter it is because, with the increased use of motor drivers, the sought thermal efficiency improvement is provided by the use of larger and more efficient combined cycle power plants, sometimes with dual and even triple pressure HRSGs. Besides, as discussed before, the pre-engineered nature of these power plants makes them much more cost-effective than custom cycles. At high fuel costs, the trade-off between capital and efficiency favours motor solutions.

CONCLUSIONS

A methodology was produced for the systematic driver and power plant selection in power-demanding industrial processes using mathematical programming. The discrete nature of gas turbines is fully considered and gas turbine drivers and pre-engineered power plants are selected from a group of candidates. An MINLP optimisation model was proposed which is able to obtain the best power system for a given economic scenario. The model is based on a superstructure approach and the designer is able to influence the final design by modifying the system constraints and economic parameters. The problem can also be formulated in an MILP fashion by linearising some of the constraints in several intervals (although it is not practical to linearise the constraints dealing with plant availability). Doing so will, without sacrificing significant accuracy, avoid the convergence problems usually associated with non-linearities and also allow more confidence in the optimality of the resulting solutions.

The MILP version of the optimisation model was applied to an LNG case study and proved to be appropriate, robust and flexible enough to produce solutions ranging from no steam at all to pure-steam systems. It was demonstrated how the optimal solution evolves and employs more thermally efficient components such as aero-derivative gas turbines, combined cycle power plants and motor drivers when faced with increasing fuel costs. The incorporation of steam turbine drivers and heat recovery steam generators made it possible to synthesise hybrid steam turbine/gas turbine solutions that, despite their increased capital cost and complexity, are able to achieve net savings and become optimal for some of the scenarios analysed, specifically in the moderate fuel cost range. No-steam solutions are still optimal when minimising capital cost only, while either hybrid gas turbine/motor or all-motor solutions proved the most cost-effective options for the higher fuel cost range. For all the cases, pure-steam systems prove to be significantly more expensive.

In the case study, the use of parallel compression was found to have a significant positive impact on the plant's economic performance. Furthermore, accounting for scheduled and unscheduled downtime directly in the objective function allowed a dramatic improvement in the total cost of the optimal series compression system. Incorporating availability in the objective function introduces a new dimension to the problem and allows considering capital/energy/availability trade-offs instead of the classic capital/energy approach. It was also shown how sparing philosophy can also be considered at designer's request to tackle the availability of electricity generation in an optimal way.

ACKNOWLEDGEMENTS

The authors would like to thank M.W. Kellogg Ltd. for their valuable technical advice and the Centre for Process Integration at the University of Manchester and the Overseas Research Student Award Scheme for the financial sponsorship of this project.

REFERENCES

Aoki, I. and Kikkawa, Y., Technical Efforts Focus on Cutting LNG Plant Costs. Oil and Gas Journal, OGJ Special, pp. 43-47, July 3rd, 1995.

Avidan, A., Messersmith, D. and Martínez, R., LNG liquefaction technologies move toward greater efficiencies, lower emissions, Oil and Gas Journal, August 19, 2002.

Bellow, E.J., Ghazai, F.P., Silverman, A.J. and Myers. S.D., Technology Advances keeping LNG Cost-Competitive. Oil and Gas Journal June 2nd, 1997.

Brooks, F.J., Gas Turbine Performance Characteristics. GE Power Systems, Technical Document GER-3567H, 2000.

Bruno, J.C., Fernandez, F., Castells, F. and Grossmann, I.E., A Rigorous MINLP Model for the Optimal Synthesis and Operation of Utility Plants. Trans IChemE, Vol. 76, Part A, March 1998.

Chou, C.C. and Shih, Y.S., A Thermodynamic Approach to the Design and Synthesis of Plant Utility Systems. *Ind. Eng. Chem. Res.*, Vol. 26, pp. 1100-1108, 1987.

Coyle, D.A., Durr, C.A. and Hill, D.K., "Cost Optimization", The Contractor's Approach. 12th International Conference and Exhibition on Liquefied Natural Gas. Perth, Australia, May 1998.

De La Vega, F.F., Durr, C.A., Hill, D.K. and Collins, C., Plant Reliability Analysis in LNG Plants. 11th International Conference on Liquefied Natural Gas. Birmingham, UK, July 1995.

Del Nogal, F.L., Townsend, D.W. and Perry, S.J., Synthesis of power systems for LNG plants, GPA Europe Spring Meeting, Bournemouth, UK, 2003.

DiNapoli, R.N. and Yost, C.C., LNG Plant Costs: Present and Future Trends. 12th International Conference and Exhibition on Liquefied Natural Gas. Perth, Australia, May 1998.

Fallaize, R.A. and Phillips, R.S., Next Generation LNG – eDrive. All Day Technical Meeting "LNG and GTL Today", Gas Processors Association Europe. February 21st, 2002.

Heckel, B.G. and Davis, F.W., Starter Motor Sizing for Large Gas Turbine (Single Shaft) Driven LNG Strings. 3rd Doha Conference on Natural Gas, Doha, Qatar, March 1999.

Iyer, R.R. and Grossmann, I.E., Synthesis and operational planning of utility systems for multiperiod operation, Computers and Chemical Engineering, Vol. 22, No. 7-8, pp 979-993, 1998.

Kleiner, F., Rausch, S. and Knabe, J., Increase power and efficiency of LNG refrigeration compressor drivers, Hydrocarbon Processing, Vol. 82, Part 1, pp 69-67, 2003.

Liu, Y.N., Lucas, C.E. and Bronfenbrenner, J.C., Optimum Design of Liquefaction Plants with Gas Turbine Drivers. 11th International Conference and Exhibition on Liquefied Natural Gas. Birmingham, U.K., July 1995.

Liu, Y.N., Lucas, C.E., Bower, R.W. and Bronfenbrenner, J.C., Reducing LNG Costs by Better Capital Utilization. 13th International Conference and

Exhibition on Liquefied Natural Gas. Seoul, Korea, May 2001.

Maia, L.O.A., Vidal de Carvalho, L.A. and Qassim, R.Y., Synthesis of Utility Systems by Simulated Annealing. Computers and Chemical Engineering, Vol. 19, No. 4, pp. 481-488, 1995.

Maia, L.O.A. and Qassim, R.Y., Synthesis of utility systems with variable demands using simulated annealing, Computers and Chamical Engineering, Vol. 21, No. 9, pp 947-950, 1997.

Manninen, J. and Zhu, X.X., Optimal Gas Turbine Integration to the Process Industries. *Ind. Eng. Chem. Res.*, Vol. 38, pp. 4317-4329, 1999.

Maréchal, F. and Kalitventzeff, B., Process Integration: Selection of the Optimal Utility System. Computers and Chemical Engineering, Vol. 22, Suppl., pp. S149-S156, 1998.

Nishio, M., Shiroko, S. and Umeda, T., A Thermodinamic Approach to Steam-Power System Design. *Ind. Engng. Chem. Process Des. Dev.*, Vol. 19, pp. 306-312, 1980.

Nishio, M., Shiroko, S. and Umeda, T., Optimal Use of Steam and Power in Chemical Plants. *Ind. Engng. Chem. Process Des. Dev.*, Vol. 21, pp. 640-646, 1982.

Papoulias, S.A. and Grossmann, I.E., A Structural Optimization Approach in Process Synthesis – I. Utility Systems. Computers and Chemical Engineering, Vol. 7, No. 6, pp. 695-706, 1983.

Peterson, N.B., Messerrsmith, D. Woodard, W. and Anderson, K., Higher Efficiency and Lower Emissions for a Base-Load LNG Plant - Are You Ready for That?, Hydrocarbon Engineering, Dec. 2001, pp. 26-28.

Richardson, F.W., Baudat, N.P., Schmidt, W.E. and Yates, D.E., Compressor and Driver Enhancements for Large LNG Plants – Look Again at Combined Cycle Options. 13th International Conference and Exhibition on Liquefied Natural Gas. Seoul, Korea, May 2001.

Shang, Z., Kokossis, A.C. and Hui, C-W., Synthesis, optimal planning and scheduling for process plant utility systems, Computers and Chemical Engineering, supplement, pp. S137-S142, 1999.

Shu, S., Christiano, F. and Harrison, M., Analysis points to electric-motor drivers for Angola LNG, Oil and Gas Journal, Oct. 7, 2002.

Varbanov, P. S. et al., Modelling and optimization of utility systems, Chemical engineering research and design, vol. 82, part 5, 2004.

Vink, K.J. and Nagelvoort, R.K., Comparison of Baseload Liquefaction Processes. 12th International Conference and Exhibition on Liquefied Natural Gas. Perth, Australia, May 1998.

Wilkendorf, F., Espuña, A. and Puigjaner, L., Minimization of the Annual Cost for Complete Utility Systems. Trans IChemE, Vol. 76, Part A, March 1998.

Zheng, J., Zhu, X.X. and Dhole, V., A Conceptual Design Approach for Power Plant Design. 6th World Congress of Chemical Engineering, California, USA, 1996.

123f Model Centric Approach for Energy Management and Optimization

Jose Desouza, Nii Asante, Piyush Shah

Energy Optimization is one of the key issues facing the process industries today. The drivers are both economic and environmental. Utilities are among the top operating expenses for manufacturers, reflecting elevated energy prices and the fact that manufacturers are using and average of between 10 and 30% more energy than necessary. In addition to capital versus energy considerations, environmental concerns now add new pressures to reduce CO2, NOx and SOx emissions.

The paper provides an overview of how site-wide utility system optimization and management can be tackled using Aspen Utilities™. Aspen Utilities addresses the key issues in the purchase, supply and usage of fuel, steam and power simultaneously within the context of an integrated utilities management objective. Aspen Utilities provides a single environment to optimize all the business processes relating to utilities management and substantially improves financial performance typically equivalent to 2 to 8% of total energy costs without requiring any plant capital investment. Additional benefits are available by using Aspen Utilities to evaluate capital investments in the utility system.

At the core of Aspen Utilities is a multi-period, mixed integer optimiser that determines the optimal configuration of equipment and distribution of loads around the utilities plant. It takes into account the cost and time taken to switch equipment on and off. The optimizer is based on a rigorous model of the utilities process as well as an exact model of all the contracts for buying and selling fuel, power and steam. The optimisation drives the system to a minimum cost operation subject to meet the current and future utilities demands within defined constraints. The system can be used in both off-line mode (e.g. by engineering staff to conduct what-if analysis) and on-line, open loop mode to provide advice, in near real time, to operations personnel.

123g Future Trends and Practical Solutions for Industrial Energy Optimization

Frank Zhu

Improving industrial energy efficiency attracts more attention from government, research, industries and public than ever. This presentation will review the past experience, current situation and future trends in improving industrial energy efficiency. The objective is to identify the needs, major issues and directions, and tools available for industrial energy management.

U.S. Department of Energy
Energy Efficiency and Renewable Energy

James E. Quinn

Plant Energy Profiler - A Software Tool for Identifying Energy Savings

AIChE Conference

April 14, 2005

Industrial Technologies Program
Energy Efficiency & Renewable Energy, Department of Energy
(james.quinn@ee.doe.gov, 202-586-5725)

U.S. Department of Energy
Energy Efficiency and Renewable Energy

Abstract

- A partnership effort between the Department of Energy Industrial Technologies Program and the AIChE was conducted to develop a DHTML-based Plant Energy Profiler (PEP) Tool. The PEP Tool enables plant energy managers to baseline energy flows in the plant; identify major energy-using equipment and unit operations; and develop a broad perspective of plant-wide energy savings opportunities. This paper will describe the unique features of the PEP Tool and demonstrate its functions and outputs.

- The PEP Tool utilizes energy end-use data on the chemicals industry from the 1998 MECS (Manufacturing Energy Consumption Survey) database to provide a default energy flow profile for the average U.S. chemical plant. Users start with this option, which requires no plant data, and then move to more detailed options which allow them to tailor the tool for their unique plant operating characteristics. The level of effort and the plant information required increases as the user moves from the first level to the last level of data input. The last level requires significant input of actual plant data and provides the closest approximation of plant energy flows. Assessments or "Scorecards" outlining plant performance are included with the last two options. The PEP Tool can be useful to plant energy managers in a number of ways. For example, the PEP Tool provides summary reports and energy maps, enabling plant engineers to regularly monitor and track energy patterns and hot spots. The PEP Tool uses qualitative assessment sheets targeting specific systems (e.g., pumps, compressors, power generators) to help identify the energy savings potential (high, medium, low) associated with individual equipment. Users are then directed to other tools and energy-saving ideas that might be pursued based on their specific plant characteristics. The PEP Tool also allows for internal energy use benchmarking among plants in the same company, and provides an individualized scorecard of plant energy performance.

- The PEP Tool was developed by Veritech, Inc. and underwent extensive beta-testing by industry experts and plant personnel prior to its official release to the public. In addition to validating the model, the beta-testing provided some excellent insights on the capabilities and ease of use of the PEP Tool with actual plant-level data.

Background – PEP Tool

- At AIChE/DOE Workshop – May 2002, industry group identified key customer needs & how US DOE could help
- Conclusion: An independent, easy-to-use tool to identify energy savings opportunities in a chemical plant
- Meetings conducted to sketch the framework of the model
- AIChE managed the project for DOE & issued a Public Solicitation
- Chemical Industry Vision 2020 Group reviewed Proposals
  - Veritech chosen as the Developer
- Veritech developed and beta tested tool
- Originally developed as excel spreadesheet; recoded in DHTML to improve useability

U.S. Department of Energy
Energy Efficiency and Renewable Energy

Basic Concepts – PEP Tool

- Provide the User with
  - Overall picture of plant energy use
  - Summary of energy cost distributions
  - Preliminary assessment & comparison
  - Identifies areas of energy efficiency improvement
  - Estimates energy cost reduction potential

Basic Features – PEP Tool

- Energy Maps
  - Electricity, Fuel & Steam
- Energy Reports
- Energy Distribution Graphs
- Potential Savings Charts
- Equipment Assessments
  - Benchmark current EE practices
  - Key equipment performance parameters
  - Energy Savings Potential

U.S. Department of Energy
Energy Efficiency and Renewable Energy

DOE's MECS Database

- Manufacturing Energy Consumption Survey
- Provides detailed energy information
  - Industry (SIC)
  - Region
  - Direct Usage (Process & Non-Process)
  - Indirect Usage
  - Import & Export
- Provided the base data for the PEP Tool

U.S. Department of Energy
Energy Efficiency and Renewable Energy

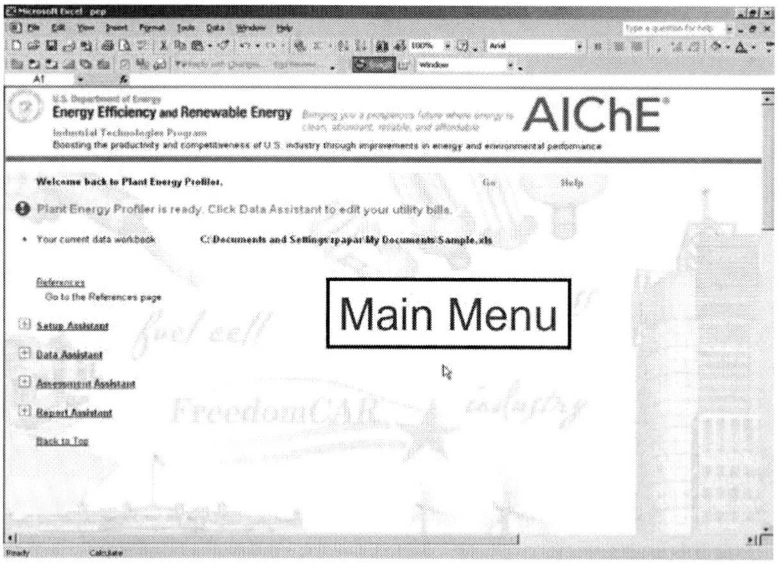

Reporting - Data Options

- US DOE Data
  - Based on MECS - Chemicals Industry
  - Will provide typical distributions
- US DOE Data with utility information
  - Actual energy usage & savings potential
- US DOE Data with utility & services info.
  - Reallocates energy usage for your plant
- My utility bills & load data
  - Accurate energy usage & savings potential

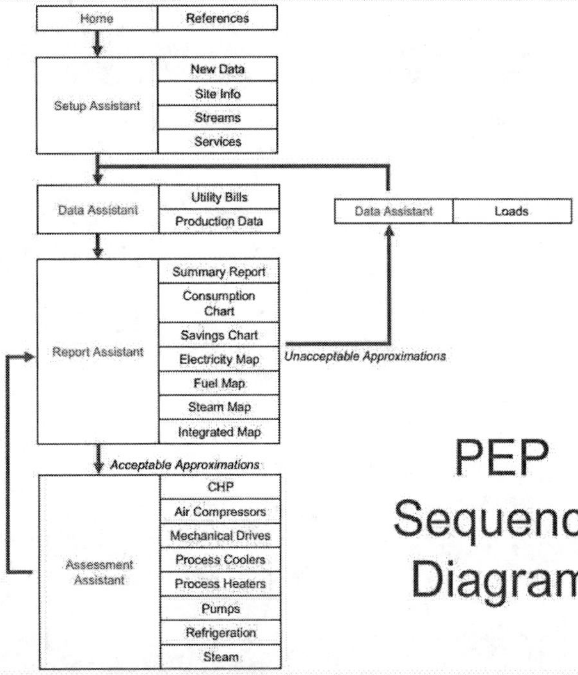

PEP Sequence Diagram

U.S. Department of Energy
Energy Efficiency and Renewable Energy

Data Entry

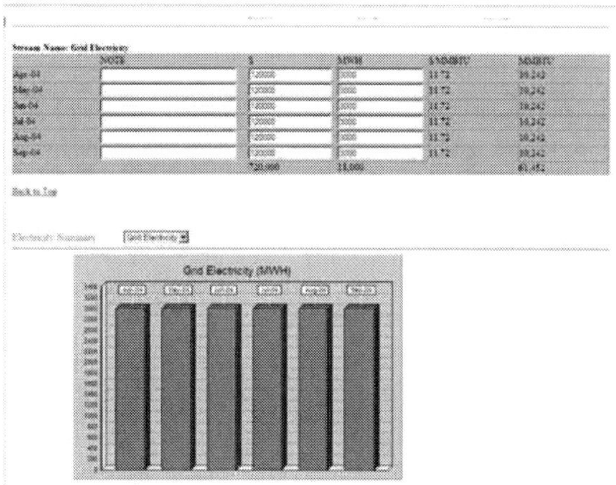

U.S. Department of Energy
Energy Efficiency and Renewable Energy

▸ Your compressed air system

Score:

1.1. Have you developed a basic block diagram of the system? (•) Yes ( ) No

1.2. Have you developed a pressure profile of your system to determine peak demand and dynamics of demand? (•) Yes ( ) No

1.3. Have you estimated total compressed air flow during different shifts? ( ) Yes (•) No

1.4. Have you measured pressure at various points in the system to determine pressure drop? ( ) Yes (•) No

1.5. Have you measured compressed air temperature at various points in the supply system? ( ) Yes (•) No

1.6. Have you estimated leak load? ( ) Yes (•) No

▸ Maintenance

2.1. Do you meet or exceed compressor and dryer manufacturer's requirements for maintenance? (•) Yes ( ) No

2.2. Do you periodically check ventilation openings to the compressor room to make sure they are free of obstructions? (•) Yes ( ) No

2.3. Do you inspect condensate drains daily? ( ) Yes (•) No

2.4. Do you periodically inspect and replace hoses that have become cracked or worn? ( ) Yes (•) No

2.5. Do you periodically inspect and replace end-use filters, check regulators, and lubricators to maintain functionality? ( ) Yes (•) No

Assessments provide checklist for finer tuni of energy savings opportunities

Assessments:
- Air compressors
- CHP
- Process Cooling
- Process Heating
- Mechanical Drives
- Pumps
- Refrigeration
- Steam

U.S. Department of Energy
Energy Efficiency and Renewable Energy

Summary Report

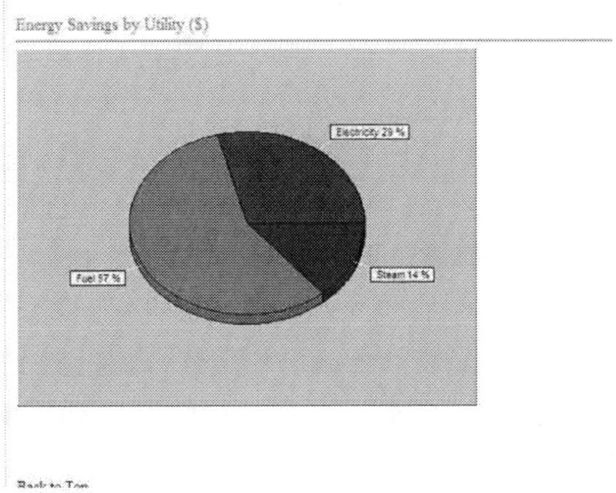

U.S. Department of Energy
Energy Efficiency and Renewable Energy

Consumption Report

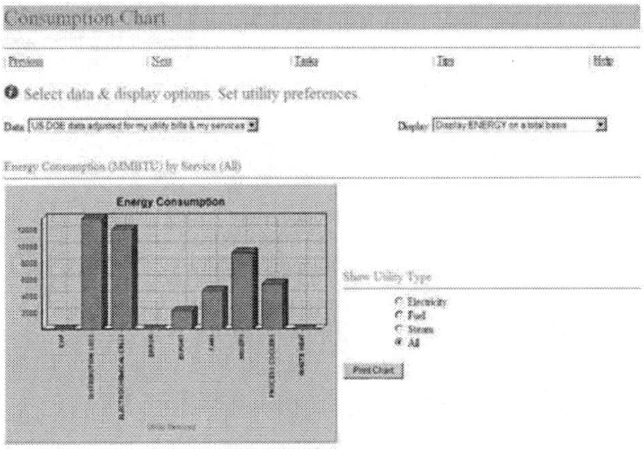

U.S. Department of Energy
Energy Efficiency and Renewable Energy

Energy Savings Report

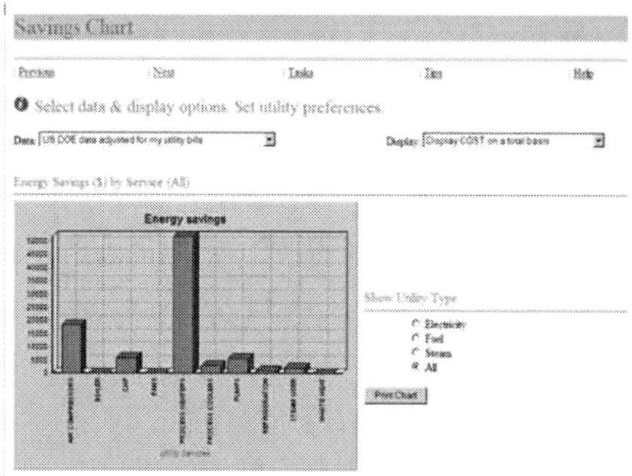

U.S. Department of Energy
Energy Efficiency and Renewable Energy

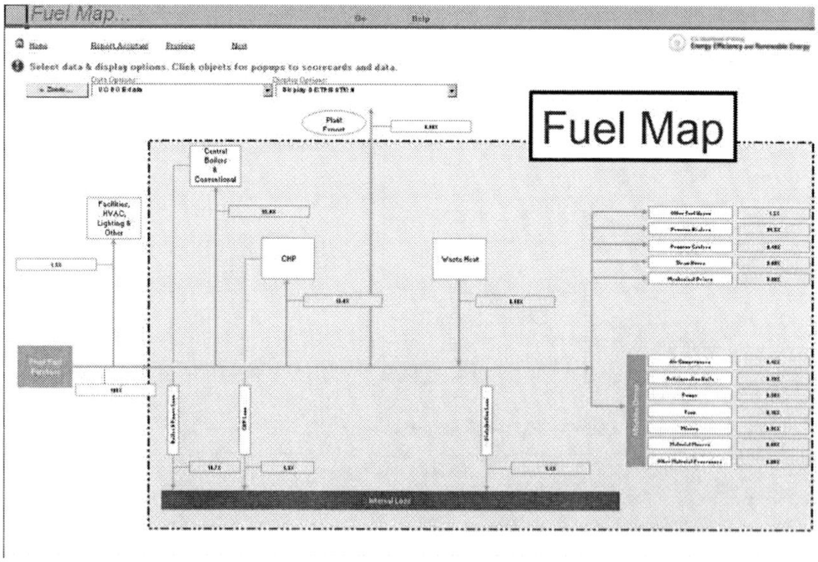

U.S. Department of Energy
Energy Efficiency and Renewable Energy

Plant Energy Profiler Tool

Use plant profile tool as entry point to using suite of tools

Integrated Tool Suite

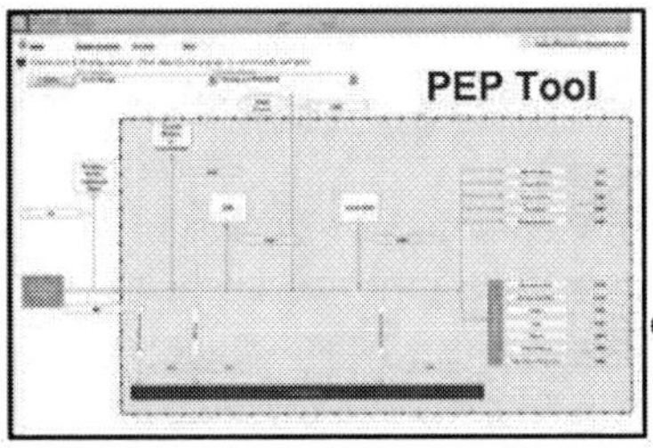

Identify Energy Savings

Opportunities by System

- 3E+
- FSAT
- PSAT
- PHAST
- SSST/SSAT
- N_xEAT
- CHP Tool
- AirMaster+
- MotorMaster+

U.S. Department of Energy
Energy Efficiency and Renewable Energy

Conclusions

- Use the PEP Tool at your plant
  - To understand energy flow
  - Identify cost impacts
  - Identify potential energy cost reduction projects
- Partner with the DOE to build Performance Benchmarks for your industry

124a BP/DPT's Efficient Gas Conversion Route

Joep J H M Font Freide and Tim Gamlin

Natural gas is an integral part of the energy market. It is also a vital feedstock for the petrochemical industry, especially for methanol and ammonia production. More recently in the refining industry, large amounts of gas are being reformed into hydrogen to upgrade fuel oils to gasoline.

Carbon monoxide and hydrogen production (syngas) from natural gas are the building blocks and key to several of BP's operations. Production of syngas is also capital cost intensive. Hence BP has always been interested in the latest technology in syngas production and since the late seventies has carried out extensive R&D in this area. From 1996 this has been carried out in collaboration with Davy Process Technology, our licensors.

The result is the Compact Reformer, which has been installed in the BP demonstration plant in Nikiski, Alaska.

Additionally the development of Fischer-Tropsch (FT) syngas conversion process to transportation fuels has been promoted by the prospect of natural gas being available at remote locations at negative or marginal cost. The FT reaction offers the possibility of manufacturing clean transportation fuels from a huge and untapped resource.

With natural gas supplies becoming more prominent and combined with environmental pressures it was important to BP to develop its FT technology. Work started in early eighties and aspects considered were product selectivity, catalyst activity and its life; all affecting the choice of reactor technology. From 1996 this again has been carried out in collaboration with Davy Process Technology, our licensors.

The results are the fixed bed reactor, which has been installed in the BP demonstration plant in Nikiski, Alaska and the development of the Advanced Jet Technology (AJT) of which a pilot plant has been installed in BP's Hull, UK facility.

The paper will address the simple flowsheet of the BP/Davy process as well as the simple exection and hence efficiency of a gas to liquids project.

sasol

reaching new frontiers

Overcoming the challenges of GTL Commercialisation

Theo Pretorius

AIChE - 2005 Spring National Meeting

Presentation Overview

sasol
reaching new frontiers

- GTL Technology: The Challenges
- Overcoming the Challenges
  - Technology Integration
  - Qatar: GTL Capital of the World
  - World Class Project Development and Implementation
- GTL and our World: Life Cycle Assessment
- Oryx GTL; Leading GTL Commercialisation from Qatar
- The Future

GTL Technology: The Challenges

sasol
reaching new frontiers

- **Low Carbon and Energy Efficiencies of earlier GTL Technology:**
  - ***High Natural Gas demand***
  - ***Large and Costly Equipment***
- **While vast improvements have been made in recent years, overall carbon efficiency is still around 75%**

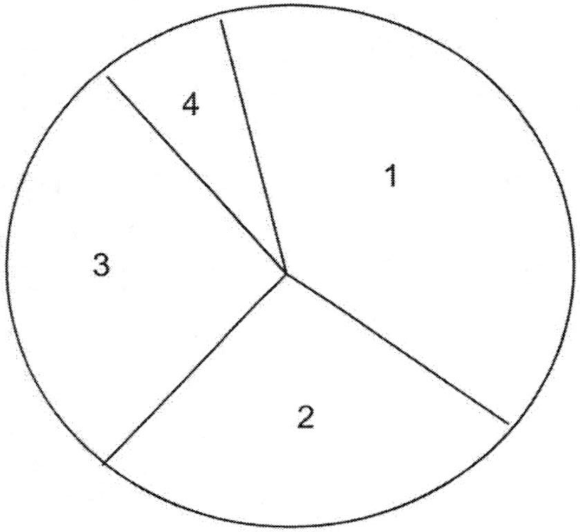

1. CO_2 from the reformer
2. Unreacted CO from FT
3. Un-recovered light and oxygenated hydrocarbons and product upgrading losses
4. CH_4 from the reformer

GTL Technology: The Challenges

sasol
reaching new frontiers

- Product Prices linked to Crude Oil Price
  - ***High Initial Capital Outlay***
  - ***Operating cost need to be competitive at low oil prices; i.e. relatively low gas price required***
- Cost of Construction in Remote Locations

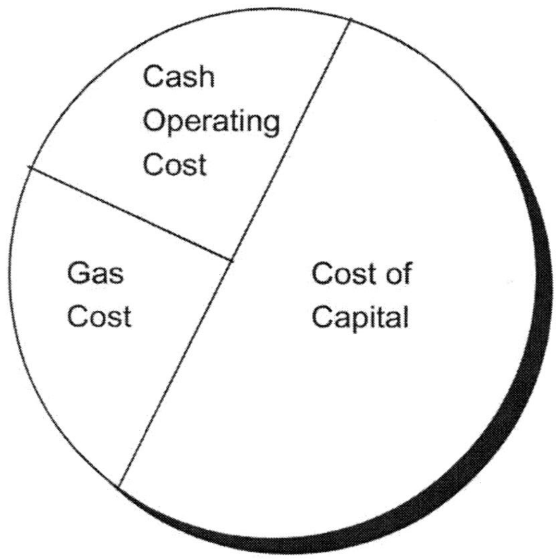

sasol
reaching new frontiers

Overcoming the Challenges:
Technology Integration

- *In order to ensure maximum process efficiency, balancing the utilities is as important as integrating the key process units*
- *Each site has an optimum point where HP steam, LP steam, fuel gas and power is in balance*
  - *Operation away from this point, reduces efficiency*
  - *Start-up needs on a remote site are an integral part of this balance*
  - *overall system design philosophy is of key importance to ensure on-line availability*
- *Sasol and its technology partners have done this and developed design tools to get there first time*

Overcoming the Challenges:
Technology Integration

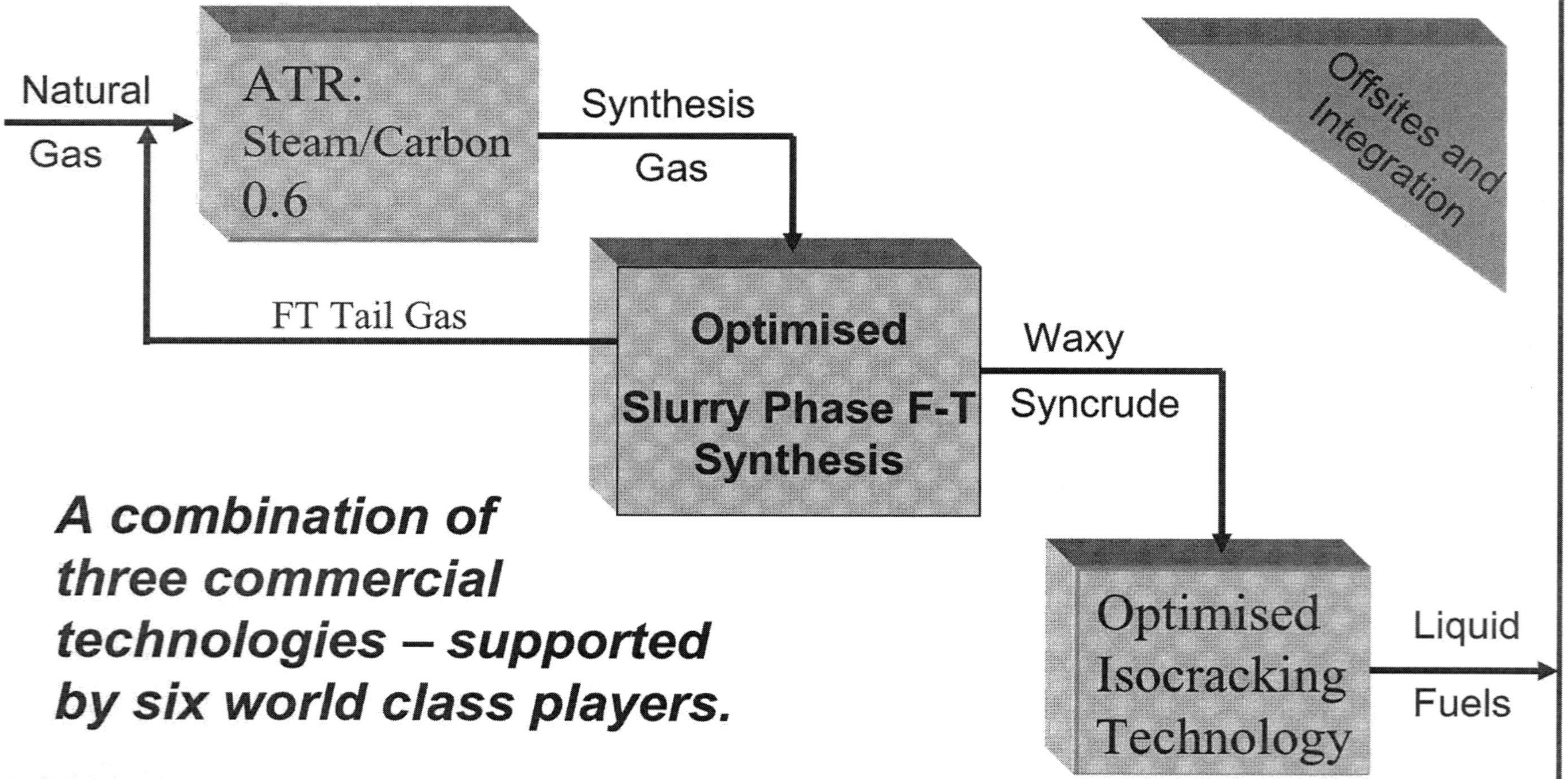

A combination of three commercial technologies – supported by six world class players.

sasol
reaching new frontiers

Overcoming the Challenges:
Qatar: GTL Capital of the World

- *Political Stability*
- *Access to Markets*
- *Large Gas Reserves (9% of world reserve)*
- *Competitive operations costs*
- *Competitive construction cost*
- *Future expansion opportunities*
- *Site synergies*
- *Modern harbour*
- *Efficient infrastructure development*
- *Supportive fiscal regime*
- *Track record with major projects*

Overcoming the Challenges:
Qatar: GTL Capital of the World

sasol
reaching new frontiers

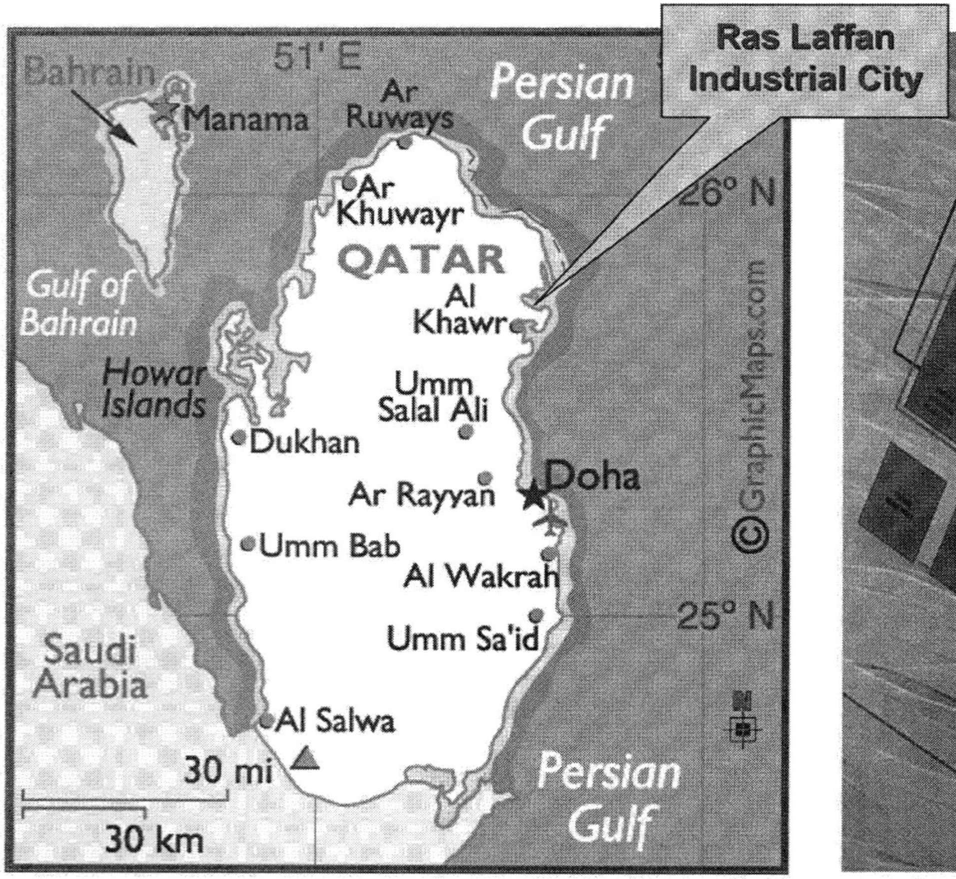

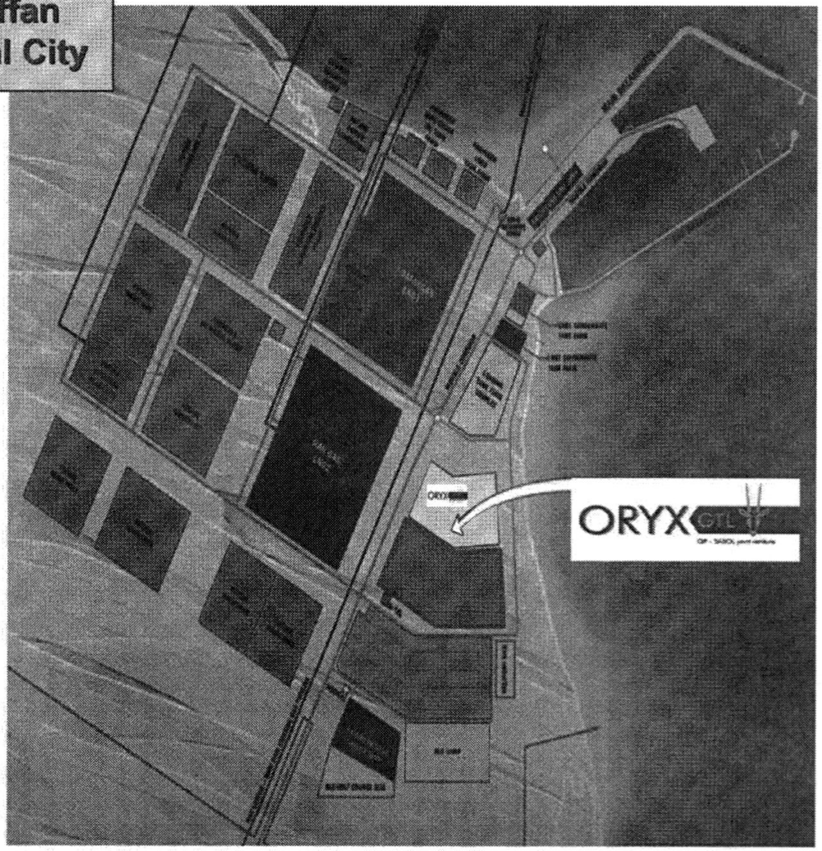

Overcoming the Challenges:
World Class Project Development and Implementation

sasol
reaching new frontiers

Front End Engineering and Development

Schedule

- Feasibility Study — 6 months
- Front End Engineering Design (FEED) — 12 months

Scope

- Full Safety Review of all Units
- 10% CAD model developed
- Full dynamic simulation of process integration
- Reliability, Availability, Maintainability (RAM) Study conducted
- ± 15% Cost Estimate

Overcoming the Challenges:

World Class Project Development and Implementation

sasol
reaching new frontiers

Technip

EPCCS Overall Schedule

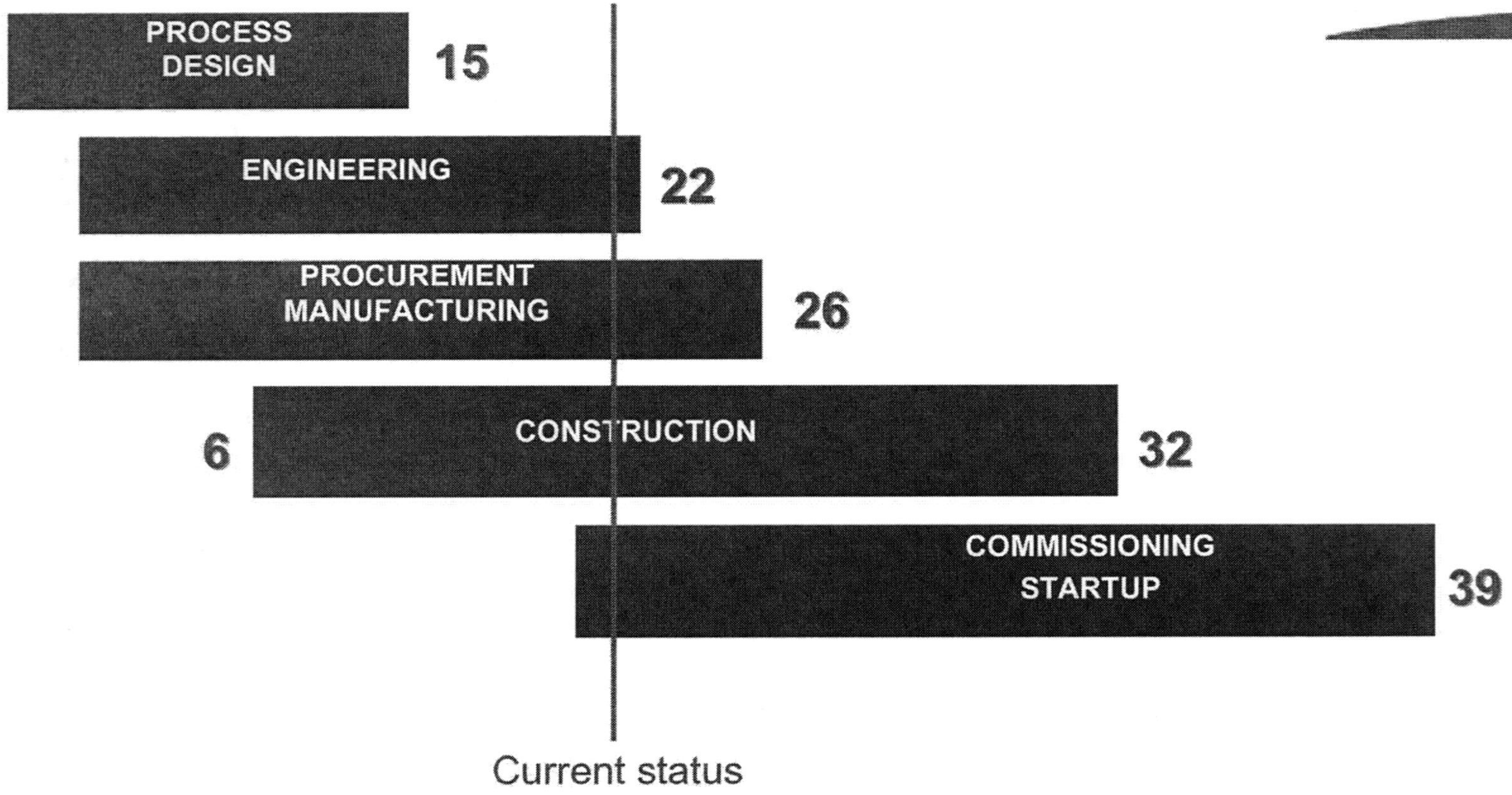

Overcoming the Challenges:
World Class Project Development and Implementation

sasol
reaching new frontiers

Technip

Project Execution – Incorporate Best Practices

- Safety is Never Negotiable
- Early Involvement of Operational Personnel – Part of FEED Team
- Project Specifications tailor made for the site
- Team Continuity
- Operation and Maintenance personnel involved in all critical reviews such as P&ID, HAZOP, Model etc.
- Conditioned Bids for critical equipment to reduce Bidder Risk
- Technology Long Leads ordered before Contract award
- Single point responsibility from Contractor including initial operation of the facility

sasol
reaching new frontiers

GTL and our World: Life Cycle Assessment

About LCA's

- LCA's are system-wide environmental assessments that consider the environmental impacts of products or technologies throughout the value chain.
- Studies based on recognised ISO standards for conducting LCA's and subject to third party critical reviews.
- The GTL LCA studies compare the underlying broader technology systems, not only the individual fuels and petrochemical produced.

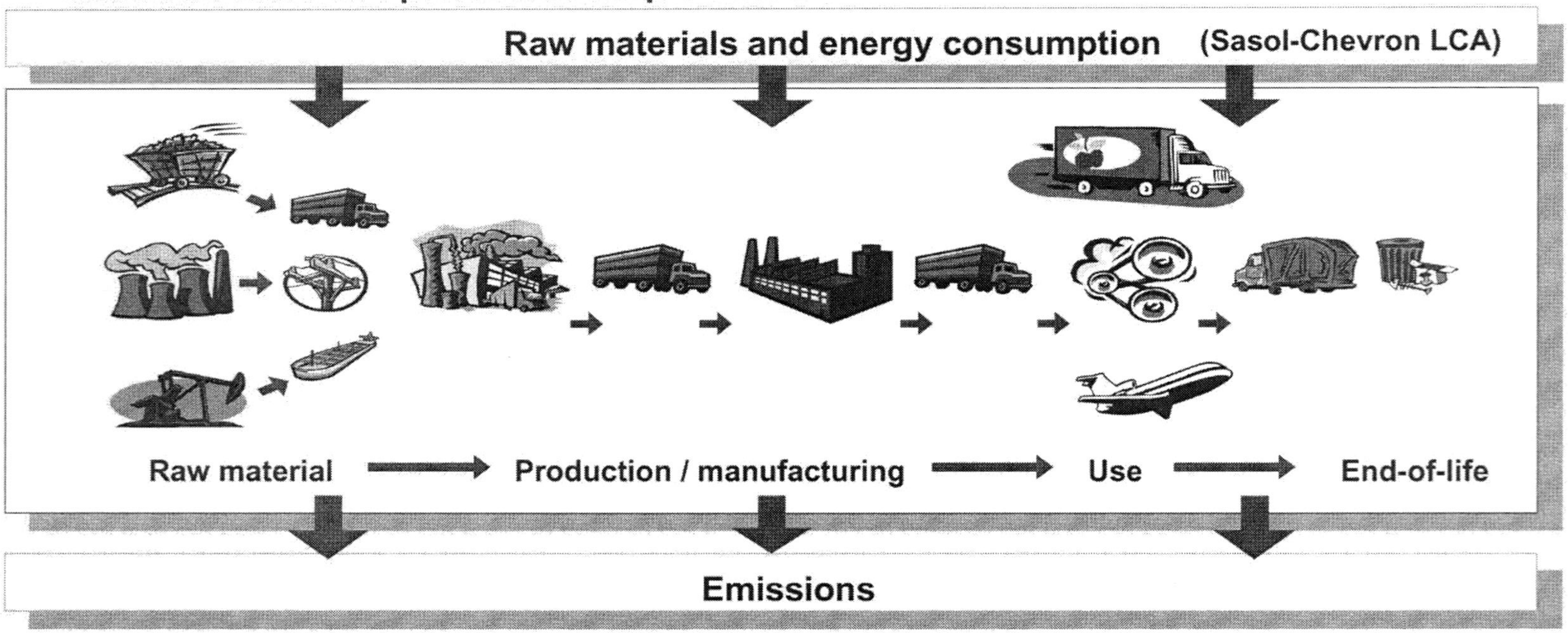

ConocoPhillips

Sasol Chevron

sasol
reaching new frontiers

GTL and our World: Life Cycle Assessment

Focus Areas of LCA's

- Greenhouse gas emissions
  - *CO_2, CH_4, N_2O → Climate change*
- Air emissions and their impacts
  - *SOx/NOx → acidification*
  - *VOC's/NOx → ozone formation*
  - *Particulates → health impacts*
- Energy efficiencies
- Resource extension / infrastructure

ConocoPhillips Sasol Chevron

sasol
reaching new frontiers

GTL and our World: Life Cycle Assessment

- *ConocoPhillips, Sasol-Chevron and Shell International Gas independently conducted Life Cycle Assessments (LCA's) on their GTL processes.*
- *Studies provide holistic views of how the technologies impact the environment caused by producing, transporting and using GTL products.*
- *Studies covered the United States, European and Asian markets*
- *In a joint effort, these three studies have been synthesised into one objective and transparent report appealing to interested stakeholders.*
- *Five Winds International, an independent LCA expert, was commissioned to prepare this report.*

ConocoPhillips Sasol Chevron

GTL and our World: *Life Cycle Assessment*

Greenhouse Gas Emissions by LCA Stage

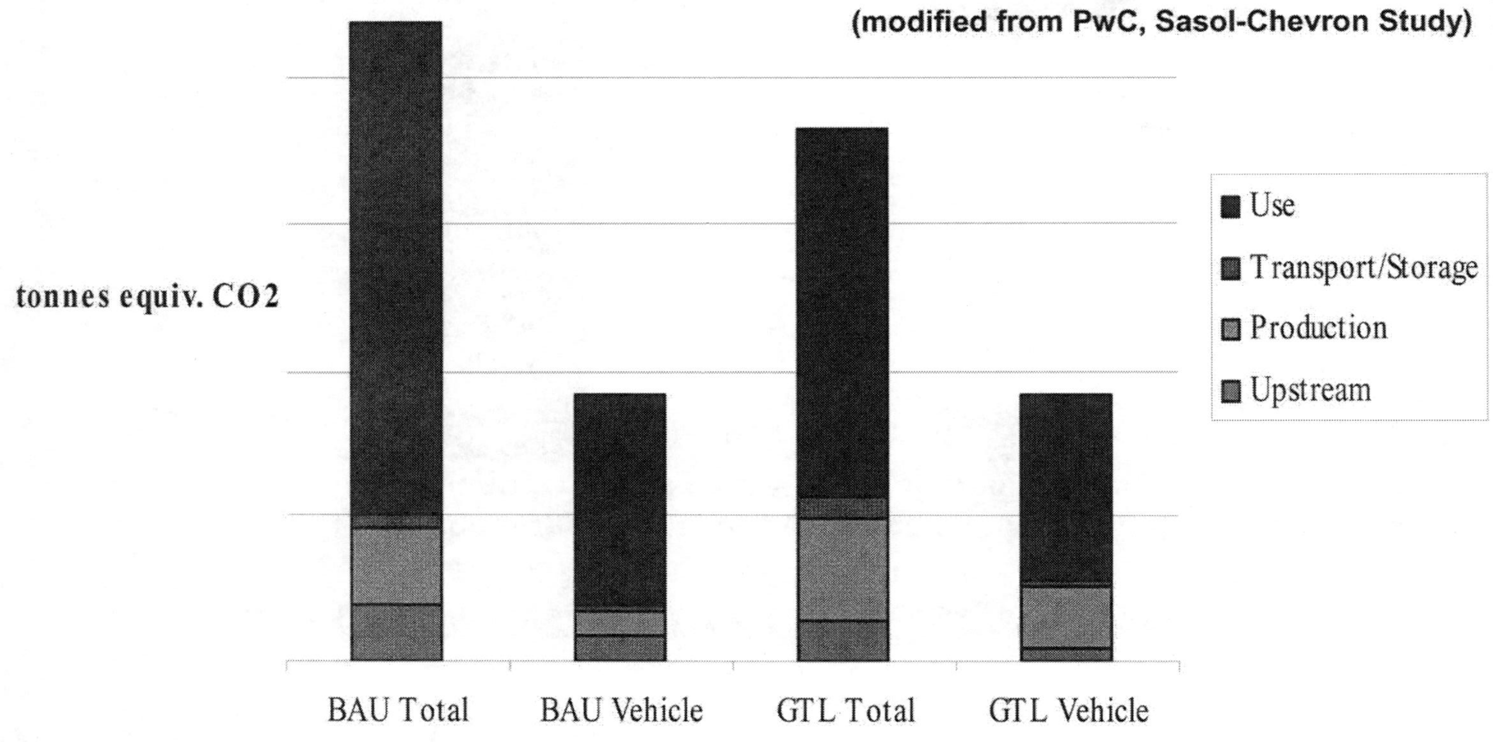

GTL and our World: Life Cycle Assessment

Photochemical Oxidant Emissions

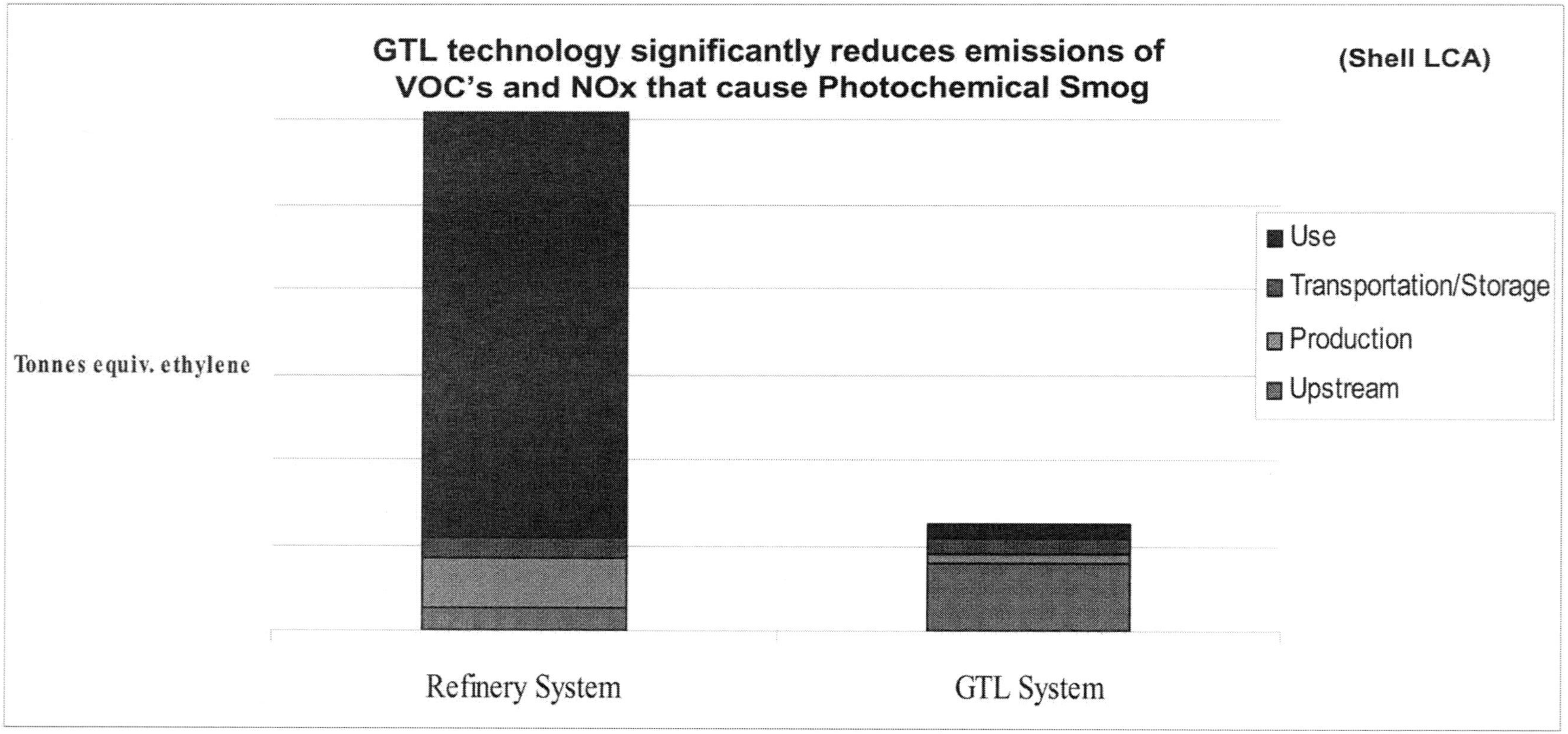

ConocoPhillips

Sasol Chevron

GTL and our World: Life Cycle Assessment

Particulate Emissions

sasol
reaching new frontiers

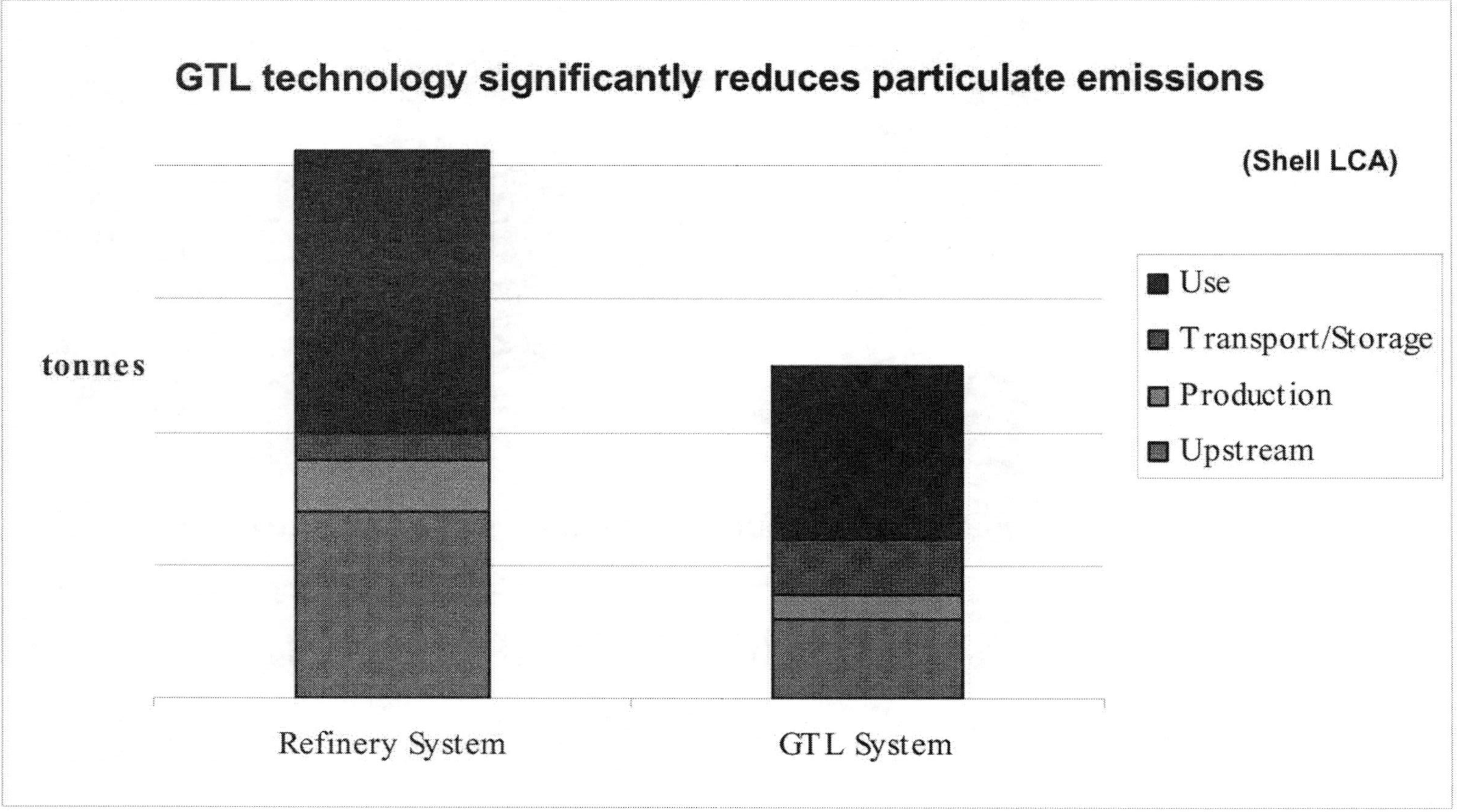

ConocoPhillips

Sasol Chevron

sasol
reaching new frontiers

GTL and our World: Life Cycle Assessment

Energy Consumption Comparisons

Energy to produce GTL is somewhat higher than conventional technologies

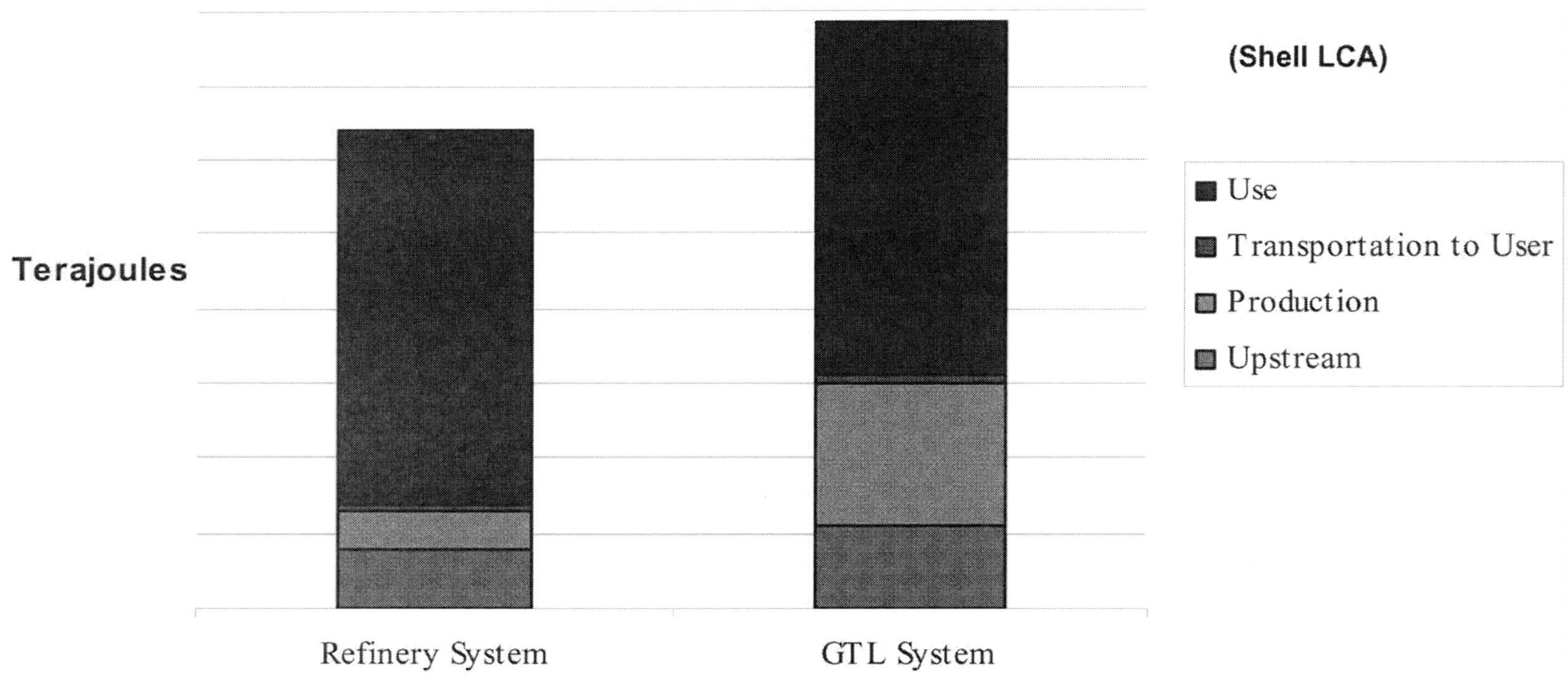

ConocoPhillips

Sasol Chevron

sasol
reaching new frontiers

GTL and our World: Life Cycle Assessment

Key Findings and Conclusions

- *The greenhouse gas impact of GTL technology is comparable to that of a modern, complex refinery, on a systems basis, and has the potential to become more favorable as technology develops.*
- *GTL technology offers advantages in terms of emissions of air pollutants (SOx, NOx, VOC's and Particulates) when compared to refinery systems.*
- *Resource availability*
    - *GTL's use of remote gas reserves extends projected life span of crude oil reserves*
    - *More energy used to produce a unit mass of GTL than conventional fuels*
    - *Fuel consumption per km driven (on a mass basis) is lower with GTL fuel than with conventional diesel.*
- *GTL technology can be used in existing infrastructure*
    - *GTL fuel can be used either directly or blended with conventional diesel and burned in conventional diesel-powered vehicles.*
    - *Fuelling infrastructure can be filled with GTL fuel without significant retrofitting or capital investment.*
    - *May enable higher distance travelled per unit mass of fuel*
- *GTL is a technology for the future with the potential to provide energy to the global market for decades by utilizing remote reserves of natural gas, the cleanest fossil fuel.*

ConocoPhillips

Sasol Chevron

ORYX GTL
QP - SASOL joint venture

sasol
reaching new frontiers

Leading GTL Commercialisation from Qatar

sasol
reaching new frontiers

Site Construction Progress

Site Construction Progress

sasol
reaching new frontiers

<u>Heavy Ends Recovery Unit</u>

sasol
reaching new frontiers

Site Construction Progress

Air Separation Unit

sasol
reaching new frontiers

Site Construction Progress

Dearators and Steam Super Heater

sasol
reaching new frontiers

Site Construction Progress

Central Piperack

sasol
reaching new frontiers

Site Construction Progress

Synthesis Gas Unit

sasol
reaching new frontiers

Site Construction Progress

Intermediate Storage Tanks

sasol
reaching new frontiers

The Future
Technology Focus

- *Catalyst Development*
- *Enhance Process Efficiency*
  - Decrease the net CO_2 selectivity of the Reformer
  - Decrease Fuel gas requirements of process fired heaters
  - Capitalise on advancements in key process steps by improved integration
- *Drive Capital Cost Down*
  - Scale is important – bigger is better
  - Operational learning will reduce the design margins
  - Integrated Projects will bring new opportunities

sasol
reaching new frontiers

The Future
Oryx is just the beginning of an exciting new industry

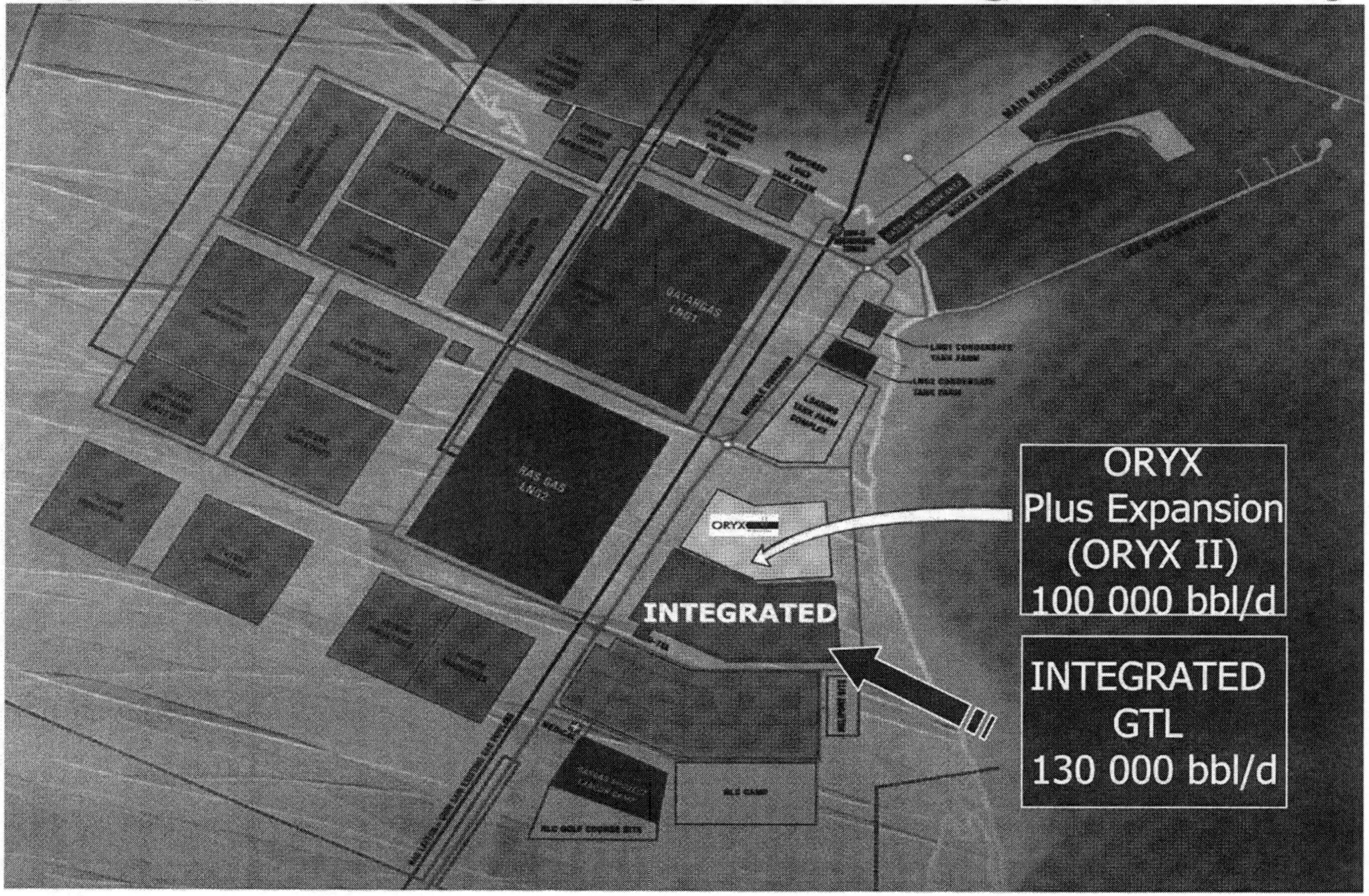

Advanced Energy Integration in Reforming Technology or Taking the Power Station out of Steam Reforming

John Swinney
Davy Process Technology Limited

Introduction

The generation of syngas by steam or auto thermal reforming is a process well known to the petrochemical industry. The process is well proven but improvements to the efficiency of the process could be made.

Two such advances in reforming technology are considered here, both of which improve the energy integration around the reformer and hence the overall plant efficiency. These are the Compact Reformer developed by BP and Davy Process Technology, and the Gas Heated Reformer developed by Johnson Matthey Catalysts.

In classical reforming the hot product gas from the reformer is cooled by raising steam. This downgrades the energy and moves it from the process to a utility which may or may not be required. Both of these alternative devices seek to utilise this energy by recycling it directly back into the reformer. This achieves two functions – it reduces the amount of energy input required to the reformer and eliminates the large quantities of by-product steam produced by conventional reforming.

The Compact Reformer, as its name suggests, is an example of process intensification that enables a large amount of reforming duty to be carried out in a relatively small unit. In the compact reformer, heat from the reformed gas is recovered into combustion air and fuel preheating. This gives high air and fuel preheat temperatures which, when combined with counter-current firing within the reformer, significantly reduces the fuel demand when compared to a conventional reformer.

The Gas Heated Reformer (GHR) operates in conjunction with an Autothermal Reformer (ATR). In the GHR, instead of firing natural gas within the reformer, the hot reformed gas from the ATR is used as the source of heat for the reforming duty. This eliminates the need for any firing on the plant, significantly improving the efficiency of the plant.

The recovery of the high grade heat by both devices is contained within a single unit which intensifies the process giving smaller and more capital efficient reforming. Also by recovering heat into the process, the production of involuntary steam is eliminated giving the design the opportunity to either remove the steam and power system from the down stream process or generate the power in a more efficient manner.

The Compact Reformer

The Compact Reformer is a small, light weight reformer that can be used in remote or offshore locations, small enough to be easily transported by road or rail but large enough to produce the quantities of syngas needed by modern plants.

The Compact Reformer consists of a number of closely packed tubes interspersed with burners to provide the heat for the reactions. This configuration means that while the chemical processes within the reformer are the same the heat transfer mechanism is significantly different to that of a conventional steam reformer. A conventional reformer has widely spaced tubes with a large volume of flue gas. The result of this is that the dominant heat transfer mechanism is radiation, with convective heat transfer from the hot flue gases accounting for less than 5% of the total heat transferred to the tubes. In the Compact Reformer this situation is reversed. The small, closely packed tubes are designed to promote convective heat transfer and the relatively small volume of flue gas reduces the potential for radiative heat transfer from the gas. Hence, about 90% of the total heat transfer is by convection.

This has resulted in a significantly more intense reforming process, which allows the same amount of reforming duty to be carried out in a much smaller unit relative to a conventional steam reformer.

To further increase process efficiency the hot process gases are used to preheat the combustion air and fuel to high temperatures. This means that three process steps involved in conventional steam reforming are incorporated into a single unit: reforming of process gas, reformed gas cooling and preheating of combustion air and fuel, further compacting the process.

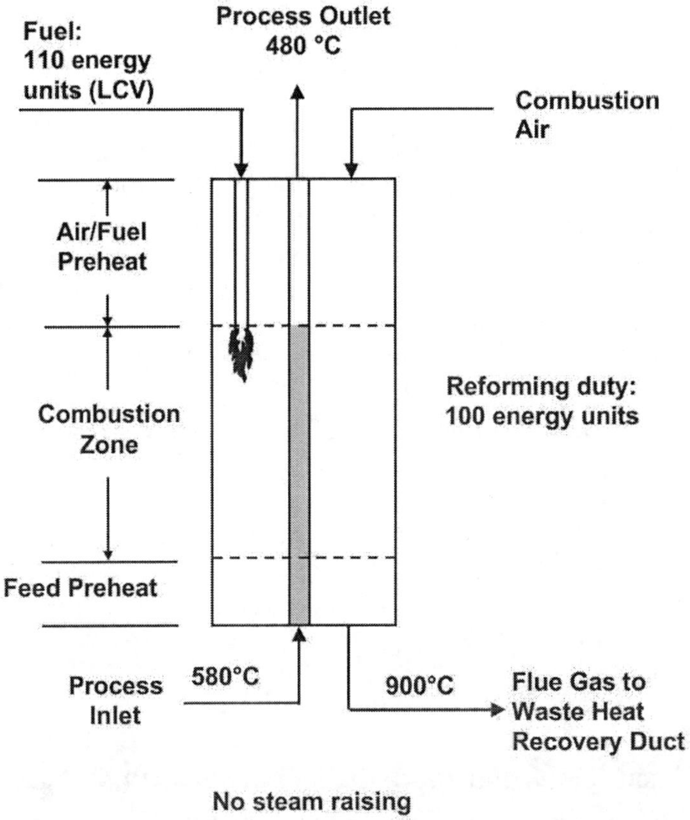

Figure 1: Configuration of the Compact Reformer

Figure 1 shows the layout of the Compact Reformer which can be considered as three distinct sections.

In the 'feed preheat' section, combustion on the shell side is complete and the combustion gases are relatively cool. Enhanced heat transfer is utilised to maximise the heat recovery from the flue gases into the process, which is also relatively cool at this stage. Reforming of the gas starts in this section.

The majority of the reforming reaction takes place in the 'combustion zone', where the temperature of the combustion gases is much higher. The catalyst bed is held at the top of the combustion zone, where the gas is disengaged from the catalyst.

The reformed gas then enters the 'air/fuel preheat' zone, where it is cooled by counter current heat exchange with combustion air. The combustion air in turn transfers heat to the fuel. Finally, the cooled reformed gas is collected in a header and leaves the reformer. The incorporation of the air/fuel preheat zone means that the reformed gas is relatively cool when it leaves the reformer, and below the metal dusting region.

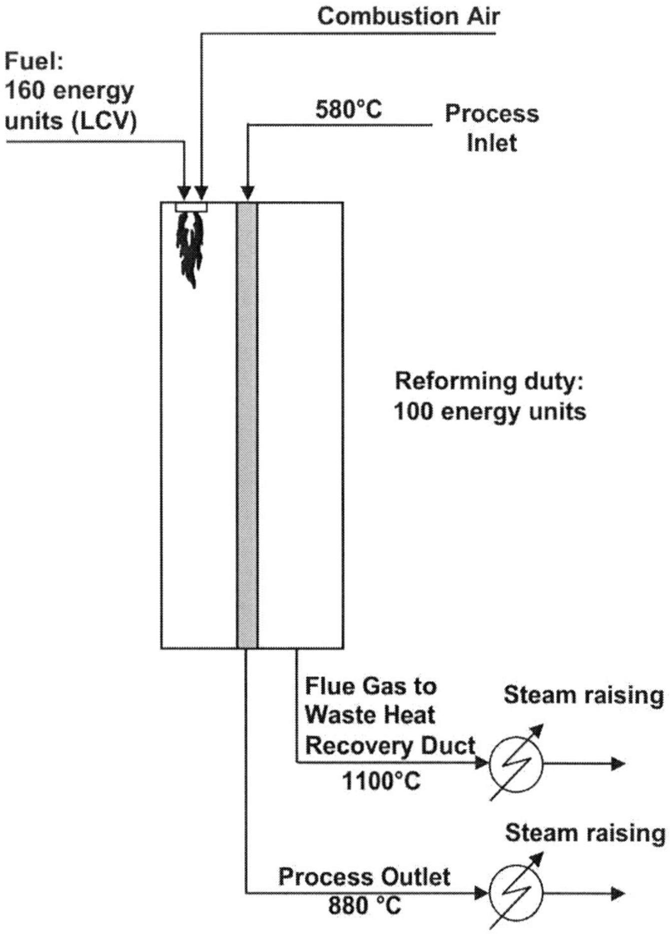

Figure 2: Conventional Steam Reformer

Traditional reformers are either co-current down fired or cross flow side fired devices. This limits the efficiency of the reformer as the flue gases must leave the reformer significantly hotter than the required process outlet temperature.

The Compact reformer is a true counter current exchanger which dramatically increases the efficiency of the process. The benefits of this can be seen by comparing Figures 1 and 2. The improved efficiency of the Compact Reformer results in approximately a 30% reduction in fuel consumption for a given reforming duty. Direct recovery of heat from the reformed gas in the Compact Reformer means that the process outlet temperature is much lower, thus eliminating the need for costly reformed gas boiler and steam system. The significantly lower flue gas temperature, coupled with the reduction in flue gas volume due to the reduced fuel consumption, means that all the heat from the flue gas can be usefully recovered into process heating, again eliminating the need for steam raising.

A full scale commercial Compact Reformer has been built and operated in Nikiski on the Kenai peninsular in Alaska. The plant is to be used to demonstrate the performance and to prove the mechanical life of the Reformer. Once this phase is completed then the Reformer will be ready for commercial operation.

Compact Reformer Performance

Based on the data from the Alaska plant and the experimental work the single Compact Reformer employed in Alaska is capable of producing approximately 26,000

Nm^3/h of syngas. This could be used to produce 150 te/day of methanol or 12,500 Nm^3/h of hydrogen.

Scaling to various capacities from relatively small hydrogen plants to world scale methanol or GtL plants can then be achieved by installing the required number of standard modules. The construction of the modules is relatively simple, and opens up the possibility for mass manufacture in a factory before transporting to site for installation. Transportation is straightforward, as a single module will fit inside a standard shipping container. Further engineering development has indicated that a version of four times the capacity can be built without compromising the modular transport concept. This concept also significantly reduces the amount of site construction required which has benefits in remote or offshore locations.

Figure 4: Compact Reformer, Nikiski

Figure 5: Installation of the Compact Reformer

The intensification of the reforming process means that significant savings in footprint and weight are achieved compared to a conventional steam reformer. On an equal capacity basis, a Compact Reformer based system occupies about 25% of the plot area of a conventional steam reformer, with a corresponding reduction in weight.

Gas Heated Reformer

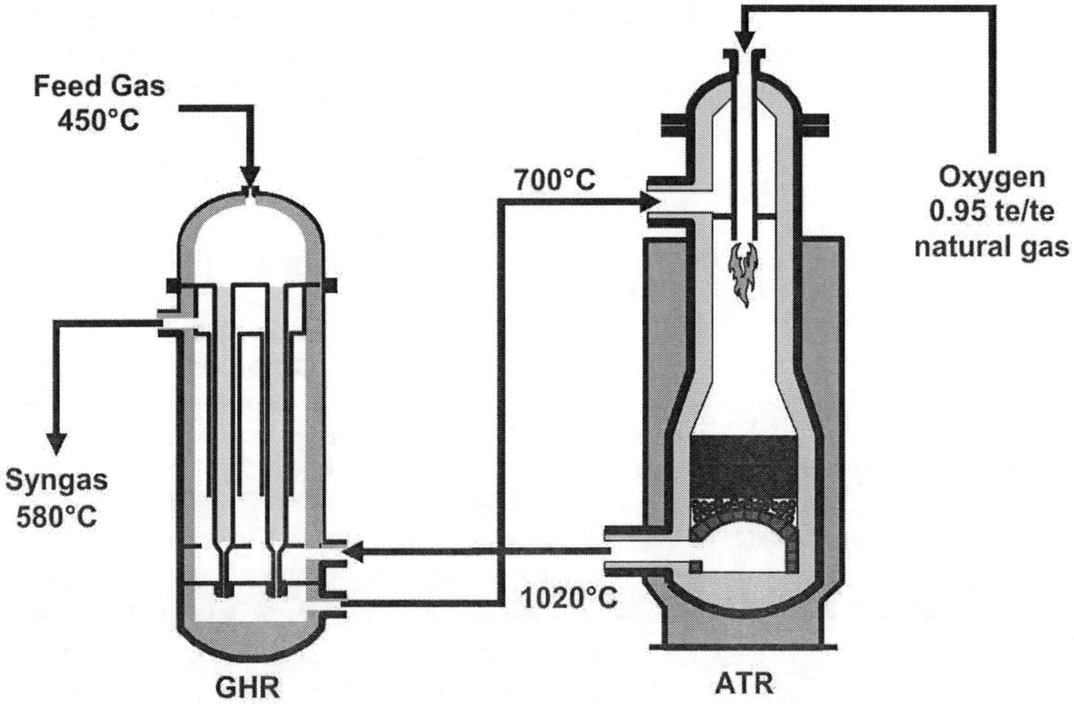

Figure 6: Diagram showing how the GHR (left) and ATR work in conjunction.

The Gas Heated Reformer (GHR) operates in conjunction with an Auto Thermal Reformer (ATR). The GHR is an interchanger combined with a reactor. Heat from the ATR (exothermic) is recycled into steam reforming (endothermic). The arrangement is illustrated in Figure 6.

Hot reformed gas from the ATR enters the shell side of the GHR and is cooled by counter-current heat exchange with feed gas passing through the tubes. The tubes are filled with reforming catalyst. The endothermic reforming reactions taking place on the catalyst provide an excellent heat sink, enabling a high heat transfer rate to be achieved.

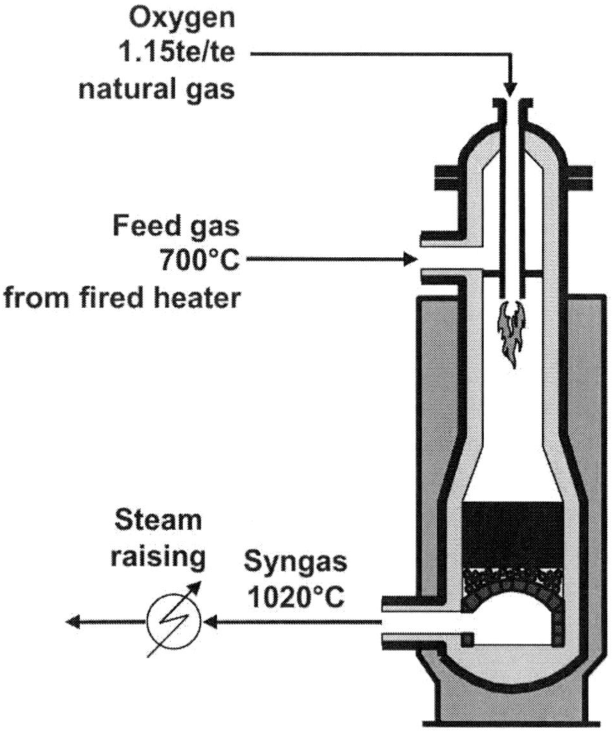

Figure 7: Autothermal Reformer

Hence, the main benefit of the GHR is that the high level heat from the reformed gas is recovered directly into the reformer. The advantages of this can be seen by comparing the GHR+ATR system in Figure 6 with the pure ATR system in Figure 7. For a given reforming duty, since part of the duty is carried out in the GHR, the oxygen consumption is reduced by 20%. The recovery of high level heat via the GHR has two further benefits: the requirement for a complex and costly steam system to cool the syngas is eliminated, and the size of the fired heater required to preheat the feed gas is reduced.

During development of the GHR, a number of complex materials and mechanical issues had to be resolved. As with the Compact Reformer, a significant amount of experimental and computer based development work went into developing the concept and turning it into an operating commercial scale facility.

The shell side of the GHR is subject to operation in an extremely corrosive environment, due to the high temperature and high partial pressure of CO which give rise to metal dusting. An extensive programme of materials development and testing was carried out to develop suitable metallurgy. The resulting design has been proven during nearly 10 years of operation on six separate GHR's. The key mechanical problem to overcome was the significant differential thermal expansion between tubes and shell. A novel solution to this was devised in the form of the patented venturi seal system. This allows each individual tube to expand freely whilst eliminating the possibility of leakage of feed gas from the tube side into reformed gas on the shell side.

The GHR is essentially an interchanger, which makes it easy to operate as it does not require any complex control systems. The main control parameter for the system is the rate of oxygen addition to the ATR, which is used to control the ATR outlet temperature.

The fact that the GHR is a "passive" device (i.e. it is not fired and has no oxygen feed) means that it is intrinsically safe. It is not possible for overheating to occur, as loss of feed would also mean loss of heat input. Even if a tube leak occurs, there is only a small pressure differential across the tubes, so all that would happen is that a small amount of partially reformed gas would leak from the tubes into the fully reformed gas on the shell side causing a slight increase in methane slip.

Scale up of the GHR to large capacities in a single unit can be relatively easily achieved by adding extra tubes. In practice the upper limit will be imposed by fabrication of the tubesheet. During design studies for world scale methanol plants, GHR designs have been prepared for single GHR's in plants of up to 6000 te/day capacity.

Conclusions

The Compact Reformer and the Gas Heated Reformer represent two different approaches to improving the energy integration around a reformer. The key to both devices is the direct recycle of high grade heat from the reformed gas back into the reformer itself. This results in improved efficiency, and eliminates the requirement for a high pressure steam system, which is the accepted method for reformed gas cooling on a conventional reforming plant – and has the effect of downgrading high level heat from the process into a utility system. Hence the goal of divorcing the reforming process from the steam system is achieved, and it is no longer necessary to have, in effect, a power station attached to the reformer. Any utilities that are required can be provided by dedicated systems.

The Compact Reformer provides a number of additional benefits over conventional reforming technology. The compact design and the consolidation of a number of processes into a single unit significantly reduces the plot area required for a given throughput. This, together with the ease of fabrication and transportability, lends itself to application offshore or in remote areas.

The Gas Heated Reformer operates in conjunction with an Autothermal Reformer and gives a significant improvement in thermal efficiency by providing a highly effective means for recovering the high level heat from the reformed gas. A significant portion of the required reforming duty is shifted from the ATR to the GHR, thus reducing the size of the ATR and Air Separation Unit for a given throughput.

124d Large Scale Production of Synthesis Gas for Methanol and GTL Plants

Kim Aasberg-Petersen, Martin R Ostberg, Per K Bakkerud

The utilisation of natural gas and associated gas for large scale production of chemicals and synthetic fuels is attracting increasing attention. Emission and flaring of these gases are in many areas prohibited for environmental reasons. Several large capacity plants utilising natural gas are either in operation or in the design or construction phase. The product in many of these plants is either methanol or synthetic diesel produced by means of the Fischer-Tropsch synthesis. The environmentally benign synthetic diesel could in future contribute to reducing emissions of harmful compounds in selected highly polluted areas.

Large scale plants producing synthetic diesel or methanol include sections for production of synthesis gas and the actual synthesis, as well as a purification section. The plant efficiencies and investment depend on both optimisation of each section and effective integration. The synthesis gas production may account for 50-60% of the investment in plant process units. The synthesis gas section is also characterised by the highest process temperatures and in many cases the largest exchange of heat and energy. Hence, technology optimisation and development often focus on especially the production of synthesis gas.

Many different technologies are proven for the production of synthesis gas. These technologies include steam reforming, thermal or catalytic partial oxidation, and Autothermal Reforming (ATR). ATR operating at low steam-to-carbon (S/C) ratio is the preferred technology for large scale production of synthesis gas. Two plants in Europe and South Africa are in commercial operation at S/C = 0.6. Other plants are in various stages of engineering and construction including a Gas-To-Liquid (GTL) plant in Qatar with a capacity of 34,000 BPD.

This presentation will mainly focus on the status and further development of the ATR technology for large scale synthesis gas production for synthesis of methanol and synthetic fuels. The influence of the selection of technology for production of synthesis gas on the energy efficiency of the overall plant will be described. The impact of main operating parameters to maximise the synthesis gas production and the carbon and energy efficiency will be highlighted. The potential of new technology from an efficiency point of view will be outlined. Finally, the possibilities for plant optimisation simultaneously with low investment and high efficiency will be discussed based on examples.

Steam Cracking and Natural Gas to Olefins: Energy Efficiency and CO_2 Emissions

Tao Ren*, Martin Patel and Kornelis Blok

Department of Science, Technology and Society,
Faculty of Chemistry and Copernicus Institute,
Utrecht University, Heidelberglaan 2, 3584 CS, The Netherlands

Prepared for presentation at the 2005 Spring National Meeting of American Institute of Chemical Engineers (AIChE) for the session "Energy Efficiency for National Gas Utilization" in Atlanta on April 10-14, 2005

*Corresponding author. Email address: t.ren@chem.uu.nl (T. Ren). Fax: +31.30.253.7601.

Acknowledgement

This PhD project is financed by Energy Research Foundation of the Netherlands (ECN) and Utrecht Energy Research Center (UCE) in Utrecht, The Netherlands. Special thanks also go to J. Hugill at ECN, P. Pujado and B. Vora at UOP LLC.

Abstract

Steam cracking for the production of light olefins (e.g. ethylene and propylene) is the single most energy consuming process in the chemical industry. Production of light olefins from natural gas is possible through different routes. This paper compares the cumulative energy use and CO_2 emissions of some of these routes (methanol-to-olefins, methanol-to-propylene and oxidative coupling of methane via ethane) with those of naphtha and ethane-based steam cracking. One conclusion is that the most efficient natural gas-to-olefins-via-methanol uses more than twice as much primary energy as state-of-the-art steam cracking to produce one ton of high value chemicals. This is largely due to high energy use in methanol production. Another conclusion is oxidative coupling of methane via ethane routes could have low CO_2 emissions if more recycling is used and less electricity is co-generated.

1 Introduction

Development of energy-efficient and low-CO_2 technologies is one of the ways to mitigate greenhouse gas emissions and to meet the target set by the Kyoto Protocol (IPCC 2001). Currently, olefin production is the single most energy consuming process in the chemical industry and its global energy use approximately 8% of the sector's total primary energy use (calorific value of final products excluded)[1]. The most common olefins are ethylene and propylene, which are mostly used to produce plastics, fibers and other chemicals. Global olefin production capacity currently is estimated to be nearly 150 million tons[2]. The conventional process for olefin production is steam cracking. The process currently accounts for approximately 2-3 EJ primary energy use (excluding calorific value of final products) and approximately 180-200 millions tons of CO_2 emissions worldwide[3]. The major feedstocks for steam cracking are naphtha and ethane. Alternatively, olefins can be produced from methane (approximately 90% of natural gas on the mass basis[4]) by means of the so-called C_1 processes (since methane contains one carbon). C_1 processes have often been hailed as the "holy grail" of the chemical industry due to their potential to replace crude oil-derived feedstocks (naphtha, gas oil, etc.) (EC-BREF 2002). In recent years, C_1 processes via methanol as intermediate have been demonstrated and commercialized. Oxidative coupling of methane is undergoing intensive R&D.

In view of these developments, the objective of this paper is to compare various C_1 processes in terms of energy use and CO_2 emissions. A few studies that estimated energy use in several C_1 processes were done in the late 80s and the early 90s (Stratton 1983; Geem 1992; Albright et al. 1992; Geerts 1991). However, an updated comparison of the cumulative energy use (see definition later) and CO_2 emissions between the most recently developed C_1 processes and steam cracking is yet missing.

In this paper, our **research question** is defined as: what are the cumulative energy use and CO_2 emissions of natural gas to olefin technologies in comparison with naphtha and ethane-based steam cracking? Energy analysis in this paper is done by first describing the processes, then characterizing them in terms of energy efficiency and finally identifying options for improvement. Energy data for both world average and state-of-the-art steam cracking are used as benchmark for the comparison with C_1 processes.

This paper is essentially a review of technical data available in the public literature. Most of the sources were found from the Internet, libraries and conferences and they belong to the following categories: journals (e.g. Oil & Gas Journal and Hydrocarbon Processing), US Patent Offices (USPTO), conferences and seminar proceedings and

---

[1] Our estimate is based on the energy data given in (IEA 2004; DOE 1998).

[2] Our estimate is based on the recent production capacity and annual growth rate reported in (Nakamura 2002; EC-BREF 2002).

[3] Our estimate is based on energy data in (IEA 2004; DOE 1998) and production data in (Nakamura 2002). See details in (Ren et al. 2004).

[4] It is reported that recoverable natural gas reserves in the world could be 5,000-6,000 EJ with a duration of 60-160 years (UNDP 2000; Rogner 1997). Additionally, there are also possibilities of deriving 840,000 EJ of methane from unconventional resources, e.g. coal-beds and hydrates (UNDP 2000; Rogner 1997).

publications by producers and licensors (e.g. UOP LLC, Lurgi, etc.). A small number of personal communications with experts also served as sources of information.

This paper focuses on the most often discussed, natural gas-related C_1 processes: three major configurations of methanol to olefins (recently commercialized) and two designs for oxidative coupling of methane via ethane (under R&D but still not commercialized). C_1 processes can convert methane to olefins either directly or indirectly. Direct conversion of methane to chemicals remains as one of the world's major scientific challenges (IFE 2003) and little publications are found so far. Indirect conversion is the focus of this paper. Also, some C_1 technologies, currently known as being far extremely energy consuming, have been excluded here. These processes require extremely high temperatures, e.g. thermal pyrolysis (1,000-1,200 °C), oxypyrolysis (850-900 °C), and chloropyrolysis of natural gas (900 °C) (Zaman 1999). This paper does not address C_1 processes based on coal and biomass. These processes are much less discussed in the current industrial research.

The content of this paper is as follows. First, two routes of C_1 processes are introduced; second, their energy use and CO_2 emissions are compared with those of steam cracking; finally, several conclusions from the comparison are drawn.

2. C_1 Processes: Methanol and Oxidative Routes

Two routes discussed in this paper are as follows:

- The methanol route involves first methanol production and then olefin synthesis (MTO or conversion of methanol to olefins). The recent development of methanol routes is summarized in Table 1. Research started approximately 20-30 years ago and two pilot plants in Norway are currently in operation. Plans for commercial-scale plants located in Africa have been announced (Table 1). Today, three major configurations have emerged and they will be explained in Section 2.1.

- The oxidation route refers to oxidative coupling of methane via ethane (OCM). It is sometimes also called partial oxidation of methane via ethane to ethylene or catalytic oxidative dimerization of methane (Moiseev et al. 2001). Table 1 lists a number of major companies involved in the R&D in OCM. The process attracted much attention from both academia and industry since the pioneering work in the early 1980s (Keller et al. 1982). The number of publications and patents reached a peak and started to decrease in 1988-1992 (Moiseev et al. 2001). The researchers in the Netherlands produced some optimistic studies in the early 1990s (Geerts et al. 1992; USPTO 1993; Geem 1992). There is neither technical nor commercial breakthrough for OCM at this moment (Hugill 2003; Pujado 2003). No pilot plants are found in recent literature.

Table 1 Developments in Methanol to Olefins and Oxidative Coupling of Methane via Ethane

Year	Companies	Processes	Capacity/status	Location	Sources
70s- 80s	Mobil, ICI, BP, Union Carbide, BASF, Hoechst, etc.	Various catalysts for converting methanol to olefins	R&D only but no commercial scale (olefin yields were very low)	Various	(EPO 1981; USPTO 1988; USPTO 1985)
1980-now	Akzo, BP, DSM, UOP LLC, ExxonMobil, etc.	Oxidative coupling of methane to olefins via ethane	R&D since 1980; approximately 20 patents were filed since 1986	Various	(Keller et al. 1982; USPTO 2000; Addroulakis et al. 2000); BP (Zaman 1999); DSM (Swanenberg 1998)
90s-now	S. C. ZECASIN	ZECASIN (Methanol to olefins)	In operation but information unavailable	Romania	(ChemSystems 2003)
90s-now	ExxonMobil, UOP LLC	Methanol to olefins and gasoline	Extensive R&D	Various	(USPTO 2003; Kunii et al. 1977; Zenz et al. 1960; Vora 2004)
2000-now	UOP LLC and Norsk Hydro	Methanol to olefins	A pilot plan with 300 ton olefins per year in operation	Norway	(Pujado et al. 2002; Barger 2001)
2002-now	Lurgi and Statoil	Methanol to propylene	Capacity of a pilot plant undisclosed; Offered for commercialization	Norway	(Hydrocarbon-Processing 2003; Koempel et al. 2003; Liebner 2002)
2003-now	UOP LLC and Eurochem	Methanol to olefins (plus polyolefin processes)	800-1,000 Kt olefins per year (under construction)[a]	Nigeria	(Alperowicz 2002)
2003-now	UOP LLC and EATCO Suez	Methanol to olefins (plus polyolefin processes)	400-550 Kt olefins per year (under construction)	Egypt	(USTDA 2000; Alperowicz 2003)

[a] The total capacity of the plants in Nigeria and Egypt together represents approximately 1% of the world olefin capacity in 2004 (Ren et al. 2004).

2.1 Methanol Route

The methanol route, as illustrated in Figure 1[1], consists of three steps: methanol production (A), conversion of methanol to olefins and gasoline (B) and product recovery and separation (C).

- **Methanol production (A)** An example of state-of-the-art methanol process is Lurgi's Mega-Methanol process. Unlike conventional methanol production in which methane is reformed to methanol with steam only, Mega-Methanol includes a step called auto-thermal reforming (or partial oxidative catalytic process) in which methane is partially oxidized to produce syngas (Lurgi 2003; Moiseev et al. 2001; Lunsford 2000). This is an exothermic reaction (approximately 5.4 GJ/t methane) and the heat released can be used for further reaction or export. A summary of the process description is as follows. Methane is first fed into a pre-reforming reactor where methane is partially reformed with steam to syngas. Then, unreformed methane is further converted into syngas with oxygen as a reforming agent in the auto-thermal reforming reactor. The reaction temperatures for steam reforming are 860-880 °C and can be up to 1000 °C. Through an exothermal synthesis process in a temperature range of 200-280 °C, syngas is converted into raw methanol (non-dewatered) (Lurgi 2003).

- **Conversion of methanol to olefins and gasoline (B)** Methanol is fed to a pre-reactor where it is partially converted into dimethyl-ether (CH_3OCH_3; commonly called DME) and water (Tabak et al. 1990; Chang 1988). The heat from methanol synthesis is generally sufficient for preheating and vaporization of methanol in the pre-reactor. Olefin synthesis occurs in the subsequent reactors (USPTO 1998; USPTO 2003). Olefin synthesis reactions are very exothermal[2]. Reactors used are similar to a fluidized-bed refinery reactor (commonly used in fluidized catalytic cracking or FCC), which is known for efficient heat transfer, temperature control and low catalyst inventory (Kvisle et al. 2001). Severity conditions are milder than those for steam cracking. Temperature is controlled to no higher than 600 °C as opposed to 750-900 °C in steam cracking. Pressure could range from 1 to 3 bar (Kvisle et al. 2001). Similar to steam cracking, high severity (higher temperature, lower pressure and short space time) favors ethylene over propylene. Another important condition here is dehydration catalyst. There are basically two major families, ZSM and SAPO. The ZSM (zeolite silicon microspores doped with metal ions such as Mn, Sb, Mg, Ba, etc.) catalyst has a shape selectivity favoring propylene and heavy hydrocarbons over ethylene (Tabak et al. 1990; Barger 2001). ZSM catalysts reportedly lead to less formation of aromatic coke and carbon oxides, than SAPO catalysts do (Barger 2001). The

---

[1] This figure is based on in the description of the EATCO's plant under construction in Egypt where UOP MTO and Lurgi's MegaMethanol processes are applied (Alperowicz 2003). We added Lurgi MTP and UOP/AutoFina's olefin conversion process to Figure 1. UOP/AutoFina's olefin conversion process is an extension of UOP MTO and in principle it can be used to upgrade byproducts from Lurgi MTP as well (Gregor 2003). ExxonMobil MTO processes have many variations, but according to the process descriptions in (USPTO 2003; Kunii et al. 1977; Zenz et al. 1960), they can also be represented by Figure 1.

[2] The total calorific values of products from methanol-to-olefin processes are smaller, or 0.5-1.5 GJ/t methanol less than that of methanol (based on yields from Table 2 and calorific values from (Chauvel et al. 1989)). Heat from the reactions at 1.7 GJ/t methanol, is released (Herz 2000). This heat, together with heat integration with methanol plants, is sufficient to provide energy needed for product separation and recovery. This explains why small amount of energy input is needed from utilities. UOP LLC claims that its process does not require a fired heater (or any energy input) (Kvisle et al. 2001; Vora 2004).

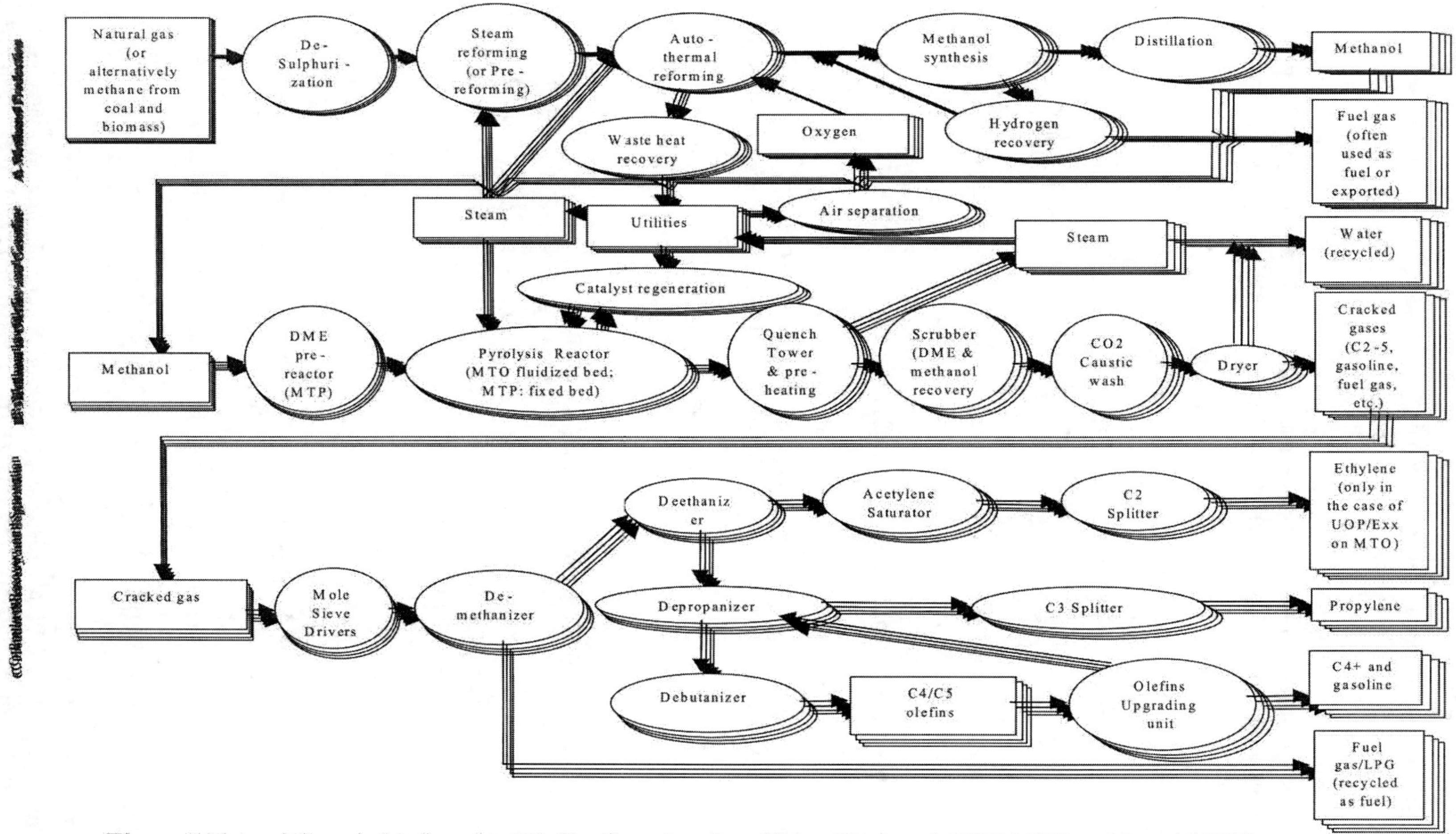

Figure 1 Natural Gas via Methanol to Olefins (based on Lurgi MegaMethanol, UOP MTO and Lurgi MTP)

Section A is Lurgi MegaMethanol process (Lurgi 2003), Section B and C represents UOP MTO (UOP 2003) and Lurgi MTP (Liebner 2002). Section C includes UOP/AutoFina Olefin Upgrading unit (Gregor 2003).

SAPO (silico-aluminophosphate molecular sieve doped with metal ions such as Mn, Ni or Co) catalyst has a shape selectivity favoring light olefins over heavy hydrocarbons. As generally true for all catalysts, both structural and chemical properties of ZSM and SAPO catalysts as well as various process parameters influence distribution of final products (Dubois et al. 2003). Dehydration catalysts are used here to convert dimethyl-ether (DME) into light olefins, heavy hydrocarbons and water.

- **Product recovery and separation (C)** The subsequent cooling, recovery and separation processes are quite similar to those of steam cracking (Pujado et al. 2002; Chauvel et al. 1989). One difference is that after the recovery and separation of C_{4-5}, the olefin upgrading (sometimes called olefin conversion) process converts C_{4-5} to ethylene and propylene (Gregor 2003). Much of the heat needed for separation and recovery processes can be supplied by heat recovered in methanol plants (USPTO 1998). Composition and yield of final products depend on catalysts and reactor configurations (dilution by steam, aromatics, etc.) and severity (temperature, space time, etc.) (Details to be explained later). High purity (97-99%), polymer grade light olefins are major products.

In this paper, three configurations of methanol to olefin processes are studied. These are: UOP LLC and Norsk Hydro's methanol to olefins process (referred to as UOP MTO in this paper), ExxonMobil's methanol to olefins and gasoline process (referred to as ExxonMobil MTO) and Lurgi and Statoil's methanol to propylene process (referred to as Lurgi MTP). The differences between UOP MTO, ExxonMobil MTO and Lurgi MTP are summarized in Table 2. The most important differences concern product yields, which are results of different reactors and catalyst applied.

2.2 **Oxidative Coupling of Methane via Ethane (OCM)**

Figure 2 shows a generic OCM process diagram with integrated electricity cogeneration and air separation for oxygen production. Table 2 shows a summary of OCM. The processes can be summarized as follows:

- **Processing of methane and oxygen** Methane is supplied from natural gas sources. Oxygen is separated from air cryogenically at the pressure of approximately six bars and at very low temperatures (commonly approximately –185 °C). Electricity is used for air separation. Depending on catalysts and process conditions, the ratios of methane and oxygen (oxygen at high purity 99%) in the feedstock can vary widely between 1:1 and 5:1 on a mass basis (USPTO 1988; Zaman 1999; Geerts et al. 1992; Huff et al. 1998). It has been suggested that this ratio be controlled at approximately 2.5:1 in order to lower the risk of explosion (USPTO 1991).

- **Oxidative coupling** A common oxidative coupling reactor design has a fluidized bed though many other configurations are also being considered (Geerts et al. 1992). In the reactor, methane (CH_4) and oxygen react over a catalyst to

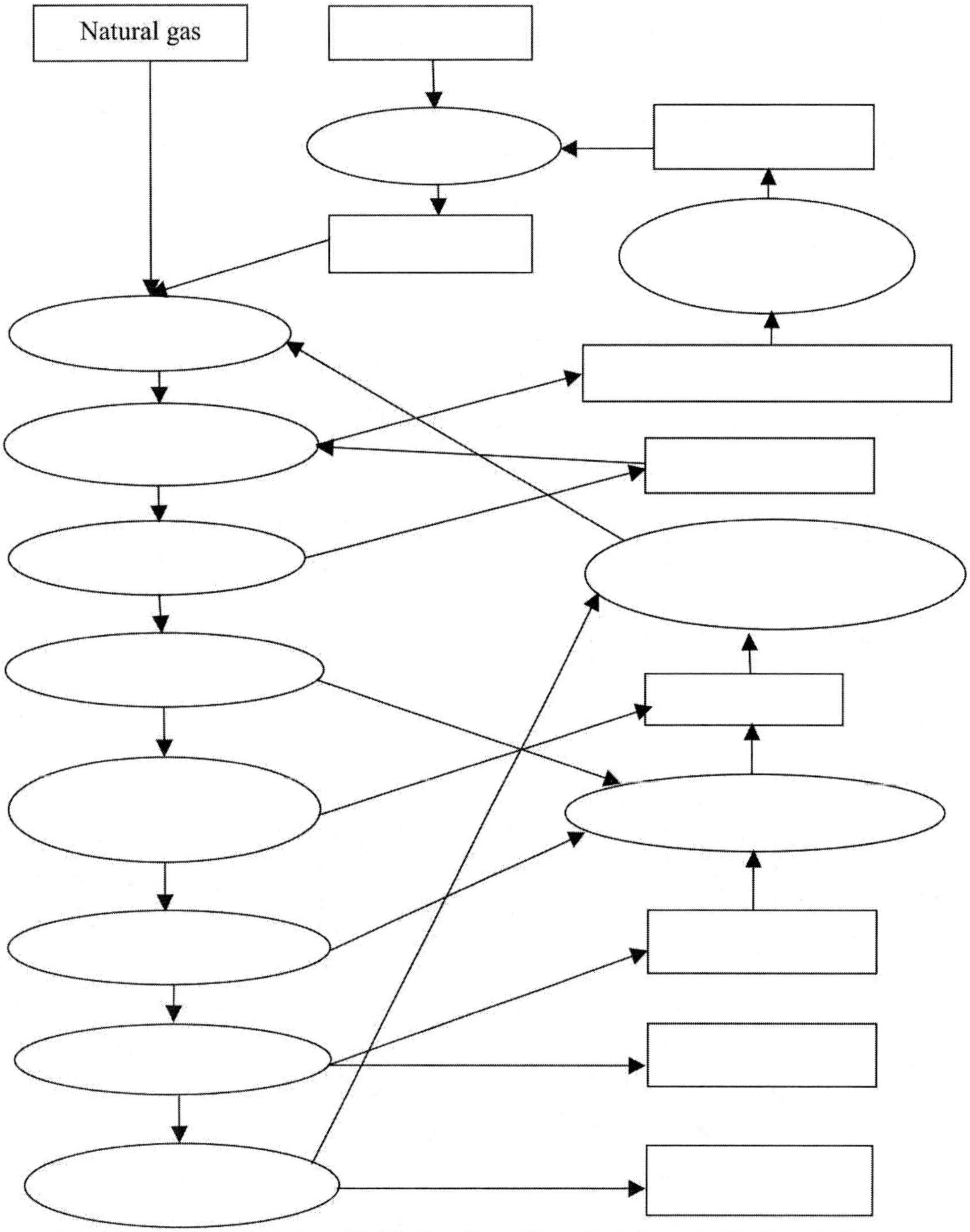

Figure 2 Process Diagram of Oxidative Coupling of Methane to Olefins via Ethane

The figure is generic. It is mostly based on (Gradassi et al. 1995; Swanenberg 1998). Electricity cogeneration is based on (Swanenberg 1998). CO_2 methanization comes from (USPTO 2000).

Table 2 Summaries of Natural Gas-to-Olefins Processes

(All yields are maximum yields and are given as the mass weights of products divided by that of natural gas)

Feedstock	Natural gas to methanol via syngas ($CH_4+H_2O \rightarrow CH_3OH$, methanol yield 145%)			Natural gas and oxygen
Processes	UOP MTO[a]	ExxonMobil MTO[b]	Lurgi MTP[c]	BP OCM[d]
Intermediate	Via methanol and DME	Via methanol and DME	Via methanol and DME	Via ethane
Desired reaction	$CH_3OH \rightarrow H_3COCH_3+H_2O$; $H_3COCH_3 \rightarrow C_2H_4+C_3H_6+H_2O$	$CH_3OH \rightarrow H_3COCH_3+ H_2O$; $H_3COCH_3 \rightarrow C_2H_4+C_3H_6+H_2O$	$CH_3OH \rightarrow H_3COCH_3+H_2O$; $H_3COCH_3 \rightarrow C_3H_6+H_2O$	$CH_4+O_2 \rightarrow C_2H_6+H_2O$; $C_2H_6+O_2 \rightarrow C_2H_4+H_2O$
Reactors	Fluidized bed	Fixed bed and fluidized bed	Fixed bed	Fixed bed
Catalysts	silico-aluminophosphate (SAPO-34 or MTO-100)	ZSM-35 (zeolite) and SAPO	ZSM	Metal oxides (e.g. MgO)
Temp. (°C)	350-525	350-500	400-450	650-850
Ethylene yield	26% w/t $C_{4\text{-}5}$ upgrading	14%	Negligible	50% per-pass yield
Propylene yield	33% w/t $C_{4\text{-}5}$ upgrading	18%	46%	7%
$C_{4\text{-}5}$ yield	9% (w/o upgrading)	Negligible	Negligible	2%
Gasoline yield	Negligible	29%	20%	Negligible
Fuel gas yield	2%	0.1%	6%	H_2 3% and others 7%
Water yield	83%	81%	81%	Water 180%; CO and CO_2 70%
Total HVCs Yield	62%	61% (45% if gasoline weighted 50%)	65% (57% if gasoline weighted 50%)	Per pass 18%

[a] UOP MTO data is based on (Kvisle et al. 2001; USPTO 2003). Olefin upgrading data is based on (Gregor 2003).
[b] ExxonMobil MTO data is based on (USPTO 2003; Keep 1999).
[c] MTP data is based on (Liebner 2002; Rothaemel et al. 2001; Koempel et al. 2003).
[d] BP OCM data is based on (Gradassi et al. 1995). No energy data is given. Its methane/oxygen ratio and ethylene yield are close to those of DSM OCM I. Its conversion rate is 22% and selectivity is 74%. DSM OCM I & II have a conversion rate of 30% and a selectivity of 50% (Swanenberg 1998).

form water and a methyl radical (CH_3). This reaction is often called partial oxidation of methane. The methyl radicals combine to form a higher alkane, mostly ethane (C_2H_6), which dehydrogenates into ethylene (C_2H_4) (USPTO 2000). The function of catalysts is to control oxygen-ions so that reactions can be kept on the desired path. Complete oxidation (rapid formation of CO_2 before the radicals link up to form ethane and ethylene) is an undesired reaction. Catalysts used are mostly oxides of alkali, alkaline earth and other rare earth metals (Zaman 1999). Hydrogen and steam are added sometimes in order to reduce coking on catalysts. After one pass, approximately 80% of the total oxygen feed by mass is consumed (USPTO 1991). The per-pass ethylene yield on a mass basis of methane is approximately 30% due to low conversion and poor selectivity to ethylene[1] (Gradassi et al. 1995; Cordi et al. 1997).

- **Compression, separation and recovery** These processes are similar to those of ethane-based steam cracking (Gradassi et al. 1995) except for the sections for water/CO_2 removal and methanization. Ethylene-containing gas streams are compressed and water is condensed. Then, the gases pass through an acid gas removal system where CO_2 is removed. Additional water is removed in a refrigeration unit and then completely removed along with CO_2. In the methanization section, CO, CO_2 and H_2 are converted to methane to increase the yield (Swanenberg 1998; USPTO 2000). From the remaining stream, ethylene, ethane, propylene and propane are respectively separated through C_2 and C_3 separation. Unprocessed methane and ethane are usually recycled, but they can also be burned as fuel along with a small amount of fuel-grade byproducts, i.e. propane, hydrogen, etc.

Given the absence of OCM pilot plants, our analysis is based on two process designs by G. Swanenberg in a DSM-sponsored feasibility study (Swanenberg 1998) (referred to as DSM OCM I & II in this paper). Both DSM OCM I & II produce ethylene, water, CO_2 and heat. Heat can be exported for steam and electricity production. Oxygen is imported. The principle differences between DSM OCM I and II are:

- Methane/oxygen ratios in DSM OCM I and II are 1:2.2 and 1.8:1 respectively.
- Recycling/methanization section is included in DSM OCM I but not included in DSM OCM II (Swanenberg 1998). In DSM OCM II all non-ethylene effluent (unconverted methane, propylene, etc.) is combusted for heat export without going through separation, recycling and methanization.
- The final outputs from DSM OCM I are ethylene (yield approximately 50% on the mass basis), electricity (approximately 20 GJ_e/t ethylene) and CO_2 (approximately 2.3 t/t ethylene[2]). The final outputs from DSM OCM II are

[1] In this paper, ethylene final yield from OCM is calculated by dividing the weight of ethylene by the mass weight of feedstock methane on the yearly basis. This definition is used in this paper to compare OCM with other processes. However, chemistry literature often reports ethylene yields on the molar basis, which is determined by conversion rate of methane and selectivity to ethylene. For example, it is said that the maximum yield of ethylene is typically close to 15-20% on the molar basis, which refers to methane conversion rate of 20-30% and selectivity to ethylene 70-80% (Gradassi et al. 1995; Pyatnitskii 2003).

[2] In the DSM OCM I process, CO_2 is captured, purified and eventually assumed to be sold if a market exists (Swanenberg 1998).

ethylene (yield approximately 15%), electricity (approximately 146 GJ_e/t ethylene) and CO_2 (approximately 16 t/t ethylene[1]).

3 Definitions

Energy indicators used in this paper are specified as follows. The cumulative energy use (based on the calculation of life-cycle energy use or net energy use[2]) is the indicator chosen in this paper. Please refer to Appendix I for another definition of energy use. The cumulative energy use is the sum of the net energy use in olefin production, which includes energy use in *feedstock production* and *olefin process*. Energy use in *feedstock production* refers to the net energy use in the production of feedstock (i.e. raw material extraction, transportation, pre-processing and production). Feedstock in this paper refers to naphtha/ethane (for steam cracking), methanol (for MTO/MTP) and oxygen/methane (for OCM processes). Energy use in *olefin processes* refers to the net energy use in the conversion of feedstock to olefins and all the subsequent processes (i.e. compression, recovery and separation).

The process heat requirement of olefin processes could be covered by the use of fuel-grade byproducts to produce heat and steam needed for reaction and separation (e.g. in the case of naphtha cracking). Alternatively, it could also be covered by external energy sources (e.g. additional fuel, steam and electricity). Oxygen is assumed to be produced by the use of electricity through air separation. Electricity is assumed to be produced by a natural gas- fired power plant, unless otherwise specified. Efficiency factors used for the generation of electricity and steam are 40%[3] and 85% (without counting energy use for fuel exploitation/production/transportation) respectively, unless otherwise specified.

All energy figures in this paper are reported in terms of primary energy, unless otherwise specified (e.g. GJ_e as electricity). In this paper, energy use is measured by specific energy consumption (or net energy use in a specific process), which is expressed in GJ/t, e.g. GJ/t feedstock, GJ/t ethylene or GJ/t high value chemicals (HVCs). For example, if specific energy consumption is expressed in GJ/t ethylene for steam cracking, it means that energy use is allocated to ethylene only and all other byproducts are hence produced "for free" in terms of energy use. This is not always the best indicator. For example, if ethane cracking is compared with naphtha cracking, it will not be fair to use the indicator "GJ/t ethylene" for comparison. Ethylene yield from ethane cracking is much higher than from naphtha cracking, but naphtha cracking also yields considerable amounts of other valuable byproducts. For this reason, in this paper the specific energy use relative to high values chemicals (GJ/t HVCs) is chosen as the primary indicator. HVCs include light olefins (ethylene, propylene and butadiene) and non-olefin chemicals. Non-olefins are mostly aromatics (and a small amount of C_{5+} hydrocarbons) in the case of steam cracking and mostly gasoline (and a small amount of butanes and C_{5+}) in the

---

[1] In the DSM OCM II process, CO_2 separation is not included and the sales of CO_2 is not assumed (Swanenberg 1998).

[2] Our definition of the cumulative energy use is equal to the difference between the total energy input and output (or the net energy use). It is not necessarily the same as the definition of process energy use (which is often used in the case of steam cracking in the literature). Using the definition of process energy use could lead to different results. See Appendix I.

[3] 40% efficiency is the current average level for power generation. This figure is assumed for the most of the processes discussed in this paper since they use very little electricity and efficiencies do not make much difference. However, OCM processes co-generate much electricity and it should be compared with a newly built natural gas power plant (efficiency 55-60%). See Section 4.3

case of C_1 processes. Backflow from naphtha steam cracking to refinery (9% yield from naphtha cracking on a mass basis) is of very low value and is not counted as HVCs[1].

In this paper, there are two approaches for defining HVCs. The first approach is to give 100% weight to the mass of olefins and non-olefins. This approach is the same as the definition of HVCs used in (EC-BREF 2002) where non-olefins are weighted with 100%. The second approach is to give 50% weight to the mass of non-olefins while the mass of light olefins is given 100% weight. The reason is that these non-olefin products are priced approximately half as much as most light olefins[2] (Gregor 2003). The second approach is useful to cross compare olefin processes that produce a variety of chemicals with very different yields and economic values. In fact, this approach only affects naphtha cracking, ExxonMobil MTO and Lurgi MTP because the yields of non-olefins from these processes are significant. For other processes, such products are negligible. Sensitivity analysis will be shown later.

In this paper, yield refers to final yield (after separation, recycling, etc.). It is defined as a percentage of desired products divided by hydrocarbon feedstock (oxygen not counted) on a mass basis, unless otherwise specified. Chemistry literature often uses per-pass yield on the molar basis. It will be noted where this definition is used in this paper. Calorific values of products are given in terms of their lower heating values (LHVs) (Neelis et al. 2003). Energy used in the production of catalyst and equipment is not included.

4 Energy and Emissions Analysis

The input data for calculating the cumulative energy use by C_1 technologies and steam cracking are listed in Table 3, 4 and 5. The following three sections (4.1, 4.2 and 4.3) will explain the assumptions, calculation processes as well as additional data that cannot be displayed in the Table 3, 4 and 5. The results of the energy and emissions analysis will be discussed in Section 4.4.

4.1 Steam Cracking

For steam cracking, naphtha and ethane are chosen as the basis for comparison because their world market share together is approximately 85% (Ren et al. 2004). Cumulative energy use in steam cracking is the sum of energy use in naphtha (or ethane) production and conversion of naphtha (or ethane) to high value chemicals. Naphtha is produced in refineries (e.g. crude oil distillation) and accounts for approximately 8% of oil refinery output on a mass basis (Hydrocarbon-Processing 2002). Ethane is typically produced through separation from natural gas (approximately 1-10% of natural gas on the mass basis in the case of wet gas[3]) and LPG or through recovery from oil refinery products (Ren et al. 2004). Energy use for naphtha and ethane production is estimated as

[1] In our calculation, the backflow has been subtracted from naphtha. The result is that the yields of high value chemicals become slight higher than what is reported as typical yields in (Chauvel et al. 1989; EC-BREF 2002). For example, in Figure 3 ethylene yield is 33% instead of a typical 30%.

[2] Light olefins market prices are $550-650/t, aromatics prices are $300-350/t and gasoline prices are $250-350/t in Western Europe based on medium prices during 1995-2003 (Gregor 2003; ChemSystems 2002; CMAI 2003; CMAI 2003).

[3] Wet gas here means the kind of natural gas that contains more ethane than dry gas, e.g. from Russia, which contains almost only methane and no ethane.

Table 3 Cumulative Energy Use in Steam Cracking Processes (calorific values of final products are excluded)

Naphtha-based steam cracking				Ethane-based steam cracking			
Steps	**Energy use per step (GJ/t product)**	**Yield (t/t)**	**GJ/t HVCs** aromatics weighted 100% (or if weighted 50%)	**Steps**	**Energy use per step (GJ/t product)**	**Yield (t/t)**	**GJ/t HVCs** (aromatics or gasoline negligible)
Naphtha production	3 GJ/t naphtha[a](state of the art)	Naphtha 8% of crude oil[b]	5 (or 5)(state of the art)	**Ethane production**	2 GJ/t ethane[c]	Ethane 1% of natural gas	3
	4 GJ/t naphtha (world average)		7 (or 7) (world average)				
Naphtha to HVCs[d]	12 GJ/t ethylene (state of the art)	Ethylene 33% and HVCs 63% of naphtha (or 60% if aromatics weighted 50%)	7 (or 7) (state of the art)	**Ethane to HVCs**	10 GJ/t ethylene (state of the art)	Ethylene 80%; HVCs 84% of ethane	9 (state of the art)
	22 GJ/t ethylene (world average)		12 (or 12) (world average)		15 GJ/t ethylene (world average)		14 (world average)
Total (GJ/t HVCs)	Not applicable		11 (or 12) (state of the art)	**Total (GJ/t HVCs)**	Not applicable		12 (state of the art)
			18 (or 19) (world average)				18 (world average)

[a] Estimation is based on (Patel 2003). Crude oil extraction, transportation and pre-processing are included.
[b] Estimation is based on (Hydrocarbon-Processing 2002).
[c] Estimation is based on (Energetics 2000). Natural gas extraction, transportation and pre-processing are included.
[d] All data on steam cracking are based on (Ren et al. 2004). Endothermicity is not included as part of cumulative energy use. Please see Appendix I for details.

approximately 7% of calorific value of crude oil and approximately 5% of calorific value of natural gas respectively[1]. Data on energy use in naphtha and ethane-based steam cracking at world average and state-of-the-art levels come from our previous study (Ren et al. 2004). These data are presented in Table 3. In this paper, energy use by state-of-the-art steam cracking technologies is the basis for the comparison with C_1 processes. The calculations for CO_2 emissions from naphtha and ethane steam cracking are estimated based on emissions factors listed in Table 6.

4.2 Methanol Routes

Cumulative energy use in the methanol routes is the sum of net energy use in methanol production and in the conversion of methanol to olefins. Energy use for methane production is assumed as 5% of calorific value of natural gas[2]. The specific energy consumption for methanol production using the latest Lurgi MegaMethanol technology (9 GJ/t methanol including energy use in oxygen production) is assumed (Lurgi 2003). This figure is slightly lower than that given by ICI (9.5-13 GJ/t methanol) (Synetix 2004). It is known that Lurgi and ICI together have approximately 90% of market share in the global chemical industry (Spath et al. 2003; Weissermel et al. 2003). The current world average (not state-of-the-art) is estimated at approximately 15 GJ/t methanol (Spath et al. 2003).

Specific energy consumption for the conversion of methanol to olefins processes (close to that of UOP MTO) is in the range of 12 to 15 GJ/t ethylene (e.g. 11.6 GJ/t ethylene in (Stratton 1983), 12.3 GJ/t ethylene in (Geem 1992) and 14.5 GJ/t ethylene in (Gradassi et al. 1995)). ExxonMobil MTO has a high energy use of 25 GJ/t ethylene due to relatively high gasoline yield (Keep 1999). Because conversion reactions are highly exothermal, most of the energy is recovered and used in separation and recovery. Energy use in olefin upgrading processes used in UOP MTO is estimated to be approximately 0.5 GJ/t $C_{4\text{-}5}$ byproducts (Gregor 2003). Our estimates for cumulative energy use (the sum of energy use in methanol production and conversion of methanol to olefins) in UOP MTO, ExxonMobil MTO and Lurgi MTP are presented in Table 4 and 5. Calculations for CO_2 emissions from UOP MTO, ExxonMobil MTO and Lurgi MTP are estimated based on emissions factors listed in Table 6 and the results are presented in Figure 4.

4.3 Oxidative Coupling of Methane (OCM)

Cumulative energy use in the OCM routes is the sum of net energy use for feedstock (i.e. methane and oxygen production) and in the conversion of feedstock to olefins. Energy use for methane production is assumed identical with that for methanol routes. Primary energy use (approximately 3 GJ/t oxygen) for production of high-purity oxygen

---

[1] The range of such energy uses for oil products is 5%-10% and for natural gas is 1-10% depending on the location and technology (Bonnekamp et al. 2001; Patel 2003). Approximately 7% and 5% for energy use for naphtha and ethane production respectively are chosen as being representative for Europe.

[2] Primary energy required in GJ/GJ delivered natural gas range widely from 0.2% to 26% depending on location (BUWAL 1995; Bonnekamp et al. 2001; Dermaut et al. 1998; Machado et al. 2001; Costanza et al. 1984). We assumed 5% since it is closed to the average in Western Europe and the US.

(99%) with the use of electricity is estimated based on (Worrell et al. 2000). The energy indicators for OCM I and II GJ/t ethylene are identical with the GJ/t HVC.

Specific energy consumption in the conversion to ethylene process[1] given in (Swanenberg 1998) is: approximately 24 GJ/t ethylene for DSM OCM I (with an average final ethylene yield of approximately 50%) and approximately 11 GJ/t ethylene for DSM OCM II (with an average final yield approximately 15%)[2]. Swanenberg took the internal value for electricity production efficiency as a reference (56% efficiency for DSM OCM I and 53% for DSM OCM II). However, we assume that the electricity would otherwise have been produced by a stand-alone, state-of-the-art power plant with 55-60% efficiency, then in the case of OCM I the specific energy consumption in the conversion to ethylene process will be 21-24 GJ/t ethylene instead of 24 GJ/t ethylene as given in (Swanenberg 1998). For DSM OCM II, specific energy consumption in the conversion to ethylene process could be 22-44 GJ/t ethylene instead of 11 GJ/t ethylene as given in (Swanenberg 1998). This high sensitivity for the latter is caused by the relatively large energy flow to electricity production. The sensitivity analysis is expressed in the error bars in Figure 3.

Energy use for feedstock production is allocated to ethylene by using the share of specific energy consumption in the sum of specific energy consumption in conversion to ethylene and energy use for electricity production (using two efficiencies 55% and 60% as just discussed). The results of allocation are given in Table 5.

In OCM I and II very little energy (less than 1 GJ/t ethylene) is lost in the reactor because in the next step heat is recovered by steam production. All net energy use occurs in the compression and separation sections (after the reactor). Due to its complex steps for separation, recycling and methanization, DSM OCM I uses twice as much as DSM OCM II. The separation section is dominated by separation of methane, ethane and ethylene (Gradassi et al. 1995), all of which are known as the most energy intensive separation steps in the case of steam cracking (Ren et al. 2004). Taking all of above into consideration, we estimate the cumulative energy uses, which are shown in Table 5 (33-36 GJ/t ethylene for DSM OCM I and 25-47 GJ/t ethylene for DSM OCM II based on the electricity production efficiency of 55-60%).

CO_2 emissions from the conversion to ethylene process are allocated in the same way as the calculation of specific energy consumption in the conversion to ethylene process. We assume that CO_2 emissions from electricity production in OCM I and II is produced by a stand-alone, state-of-the-art, natural gas-fired power plant with efficiency of 55-60% (CO_2 0.102 t/GJ_e and 0.093 t/GJ_e respectively). These emissions are then deducted from the total emissions from OCM I and II (see Section 2.2). CO_2 emissions from feedstock

---

[1] Process energy use in OCM olefin production is relatively small if compared with steam cracking (see others in Table 3 and 5) (Ren et al. 2004). However, it is not surprising because OCM is very exothermal: the heat emitted from oxidation is 15-20 GJ/t methane (Cordi et al. 1997). The total calorific value of products from OCM is much smaller than that of methane due to the conversion of methane to heat.

[2] Our estimates for DSM OCM I and II is much higher than two early studies on different configurations of oxidative coupling (Geem 1992; Geerts et al. 1992). These two studies estimated energy use (fuels and electricity) as approximately 5 GJ/t ethylene in primary energy terms. A final ethylene yield of 70% (much higher than those of DSM OCM I and II) after recycling ethane and methane is assumed. Energy use in methane and oxygen production was not excluded.

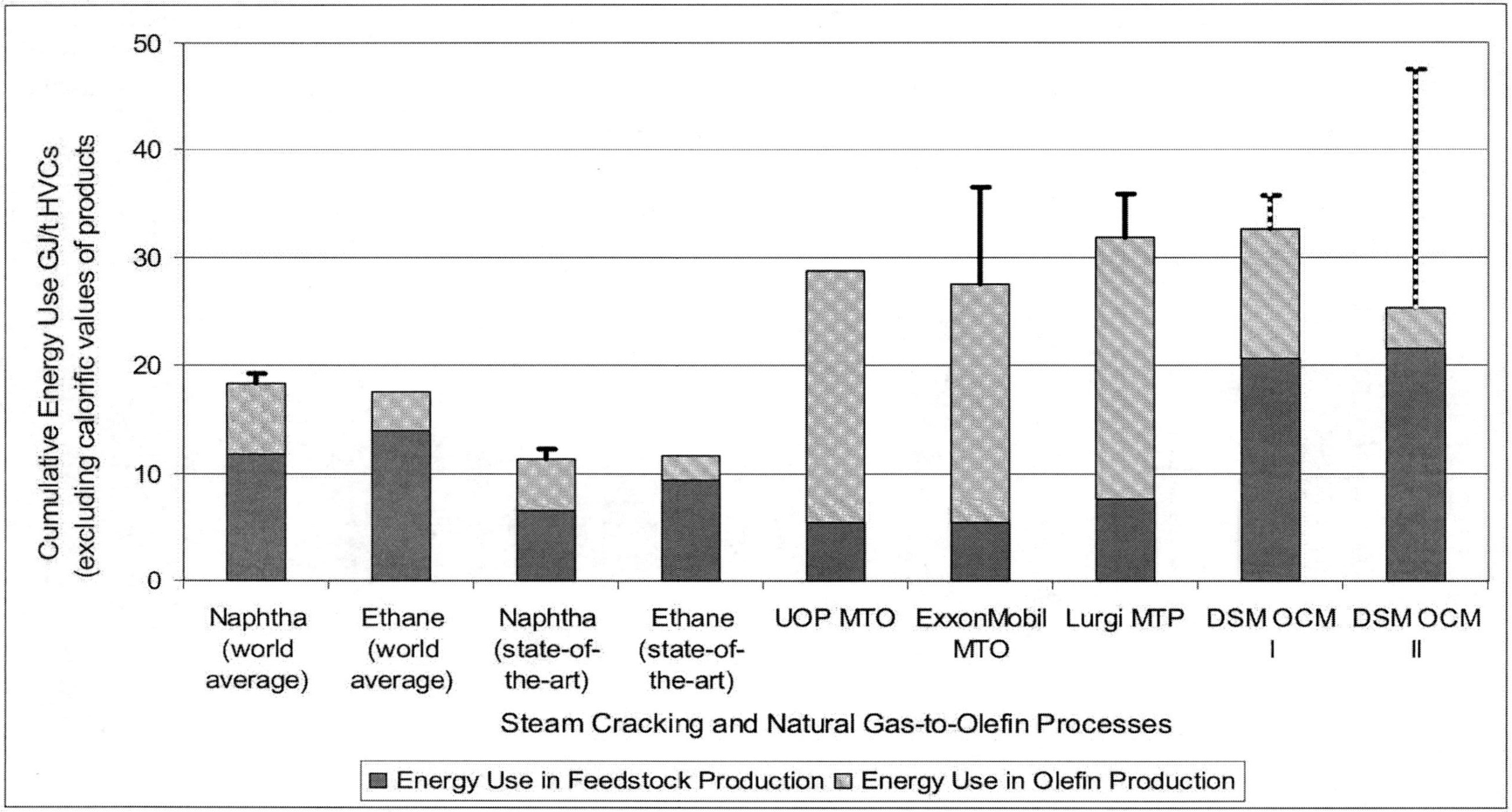

Figure 3 Comparison of Cumulative Energy Use by Steam Cracking and C_1 Technologies with Sensitivity Analysis for Different Definitions of High Values Chemicals and Power Generation Efficiencies (calorific values of products excluded)

Data come from Table 3, 4 and 5. HVCs have calorific values of approximately 42-44 GJ/t. The solid errors bars for naphtha cracking, ExxonMobil MTO and Lurgi MTP are the results if 50% (weight) of aromatics and gasoline is counted as HVCs (whose calorific values are approximately 50 GJ/t). See details in Section 3 and Appendix I. The dotted error bars for OCM I & II are the results if power generation efficiency of 60% is used. The results by using power generation efficiency of 55% are the default. See detailed explanation in Table 3, 4 and 5.

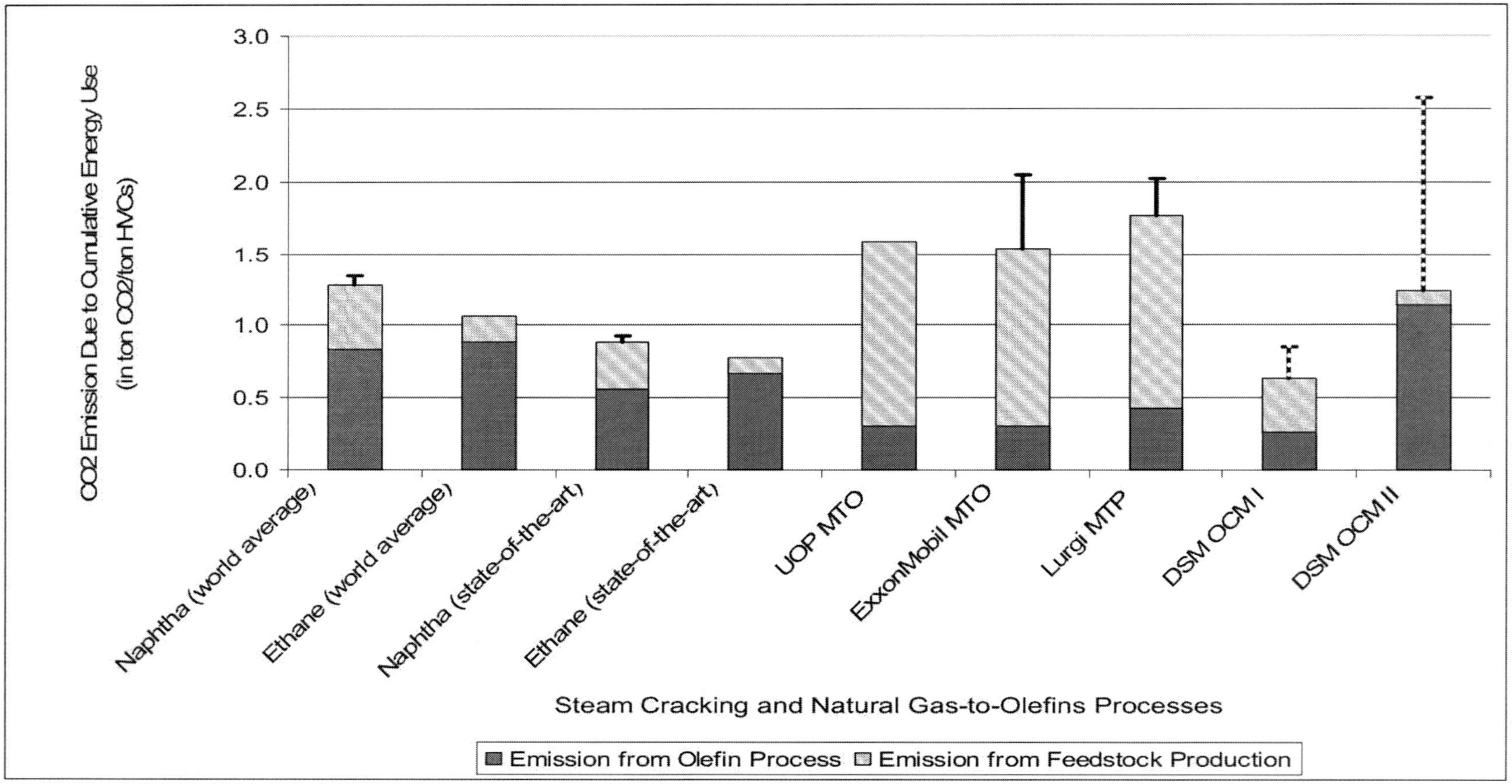

Figure 4 Comparison of Estimated CO_2 emissions by Steam Cracking and C_1 Technologies with Sensitivity Indication for Different Definitions of High Values Chemicals and Efficiencies

Data is based on Table 3-6. The error bars represent the same sensitivity analysis as those in Figure 3. Please note that CO_2 emissions from DSM OCM I processes are very low as the result of partial oxidative, exothermic reaction. See Table 6 for emission factors and explanations.

production are allocated by the same way as the allocation of energy use in feedstock production.

The results of CO_2 allocation are shown in Table 6 (emissions factors) and in Figure 4 (emissions per ton of ethylene). Please note two features. First, DSM OCM I has relatively low emissions. This is mainly due to the partial oxidative, exothermic reactions in which thermal energy is released without producing CO_2 (see the OCM reaction equation in Table 2). This overcompensates the joint effect of undesired side reaction and energy requirement for separation that lead to CO_2 emissions. Second, the CO_2 emissions from OCM I are lower than from OCM II while its energy use (in Figure 3) is higher than OCM II. OCM I uses more energy because it has more recycling and less electricity co-generation than OCM II does. Its emissions are lower because it combusts less fuel for electricity cogeneration than OCM II does.

4.4 Results of Energy and Emissions Analysis

The results of energy and emissions analysis in terms of HVCs are shown in Figure 3 and Figure 4. All energy uses and CO_2 emissions by C_1 processes are higher than those of naphtha and ethane cracking. Among the methanol routes, ExxonMobil MTO and UOP MTO are the most efficient, but their energy uses are still approximately 150% higher than that of state-of-the-art naphtha cracking. Similarly, CO_2 emissions by all methanol routes are much higher than state-of-the-art ethane cracking. DSM OCM I and II have very different energy use and CO_2 emissions due to different designs for recycling and electricity co-generation.

5 Further Improvements

5.1 Methanol Routes

Compared with the energy use in steam cracking, methanol routes have three features that are interesting to be discussed here: *methanol, dehydration catalysts and product treatment.* First, the energy required for methanol production is much higher than that for steam cracking for naphtha and ethane production. For this reason, the cumulative energy use of methanol routes is already higher than that of steam cracking. Methanol as a feedstock is used because it is currently the most economically and technologically proven choice among all gas-to-liquid technologies. While there are currently no technological breakthroughs in sight that can make methanol routes more energy efficient than steam cracking, in theory much room for energy saving in methanol production exists. The theoretical minimum exergy consumption of an overall methane-to-methanol reaction is approximately 3 GJ/t methanol (Weissermel et al. 2003; Machida et al. 1998) (for comparison: 9 GJ/t methanol according to Lurgi mentioned earlier). If it were possible to reach this theoretical value, then methanol routes would be as efficient as the current naphtha steam cracking. Converting methane to syngas is currently done through steam reforming. Steam reforming requires very high temperatures (at 800-1000 ^{o}C) and it is highly endothermic[1] (approximately 30 GJ/t methane is combusted to

[1] The overall methanol production is exothermal. Calorific value of feedstock methane is approximately 30% higher than that of methanol.

Table 4 Cumulative Energy Use in Natural Gas-to-Olefins Processes (Part I)

Natural gas-to-Olefins (UOP MTO)				Natural gas-to-Olefins (ExxonMobil)			
Steps	**Energy Use per step (GJ/t product)**	**Yield (t/t)**	**GJ/t HVCs** (aromatics or gasoline negligible)	**Steps**	**Energy use per step (GJ/t product)**	**Yield (t/t)**	**GJ/t HVCs** gasoline weighted 100%) (or if gasoline weighted 50%)
Natural gas production and Methanol production[a]	10 GJ/t methanol	Methanol yield 145% of natural gas[b]	23	**Natural gas production, Methanol production**	10 GJ/t methanol	Methanol yield 145% of natural gas	23 (or 30)
Methanol to HVCs	13 GJ/t ethylene[c]	Ethylene 26%; HVCs 62% of natural gas	5	**Methanol to HVCs**	25 GJ/t ethylene[d]	HVCs 61% of natural gas (or 45% if gasoline weighted 50%)	5 (or 7)
Total (GJ/HVCs)	Not applicable		29	**Total (GJ/HVCs)**	Not applicable		28 (or 37)

[a] Methanol data is based on (Lurgi 2003). Approximately 2 GJ/t natural gas is used for natural gas production (Patel 2003). Natural gas extraction, transportation and separation are included.
[b] Methanol yield is expressed by the weight of methanol divided by that of natural gas, which is larger than 100% because methanol contains oxygen (Spath et al. 2003).
[c] Estimation is based on (USPTO 2002; USPTO 2003). Energy use in olefin upgrading approximately 0.5 GJ/t $C_{4\text{-}5}$ has been accounted for (Gregor 2003).
[d] Estimation is based on (Keep 1999).

Table 5 Cumulative Energy Use in Natural Gas-to-Olefins Processes (Part II)

Natural gas-to-Olefins (Lurgi MTP)				Natural gas-to-Olefins (DSM OCM I and II)			
Steps	**Energy Use per step (GJ/t product)**	**Yield (t/t)**	**GJ/t HVCs** gasoline weighted 100% (or weighted 50%)	**Steps**	**Energy Use per step (GJ/t product)**	**Yield (t/t) (all divided by natural gas)**	**GJ/t HVCs** (aromatics or gasoline negligible)
Natural gas production, Methanol production	10 GJ/t methanol	Methanol yield 145% of natural gas	25 (or 28)	**Natural gas production**	2 GJ/t natural gas	Not applicable	3 (DSM I) 2 (DSM II)
				Oxygen production	3 GJ/t oxygen[a]		9 (DSM I) 2 (DSM II)
Methanol to HVCs	10 GJ/t propylene[b]	Propylene 46%; HVCs 65% of natural gas (or 57% if gasoline weighted 50%)	8 (or 9)	**Conversion to ethylene** (electricity production efficiency of 55-60%)	21-22 GJ/t ethylene[c]	HVC 50% (DSM I)	21-22 (DSM I)
					22-44 GJ/t ethylene	HVC 15% (DSM II)	22-44 (DSM II)
Total (GJ/HVCs)	Not applicable		32 (or 36)	**Total (GJ/HVCs)**	Not applicable		33-36 (DSM I), 25-47 (DSM II)

[a] Estimation is based on (IKARUS 1994).

[b] Estimation is based on (Koempel et al. 2003; Spath et al. 2003).

[c] Estimation for DSM OCM I & II is based on (Swanenberg 1998) but with an efficiency of 55% for power generation. Energy use in compression, recovery and separation is included and is clearly higher in the case of OCM I than that in the case of OCM II.

Table 6 Emissions Factors for Steam Cracking and Natural Gas-to-Olefins (ton CO_2 per GJ Energy Consumed)

	Naphtha based steam cracking	Ethane based steam cracking	Natural gas-to-Olefins				
			UOP MTO	ExxonMobil MTO	Lurgi MTP	DSM OCM I	DSM OCM II
Feedstock production	0.069 (combustion of refinery gas for naphtha production)[a]	0.056 (combustion of natural gas for ethane production)[b]	0.056 (combustion of natural gas for methane production)[c]			0.056 (combustion of natural gas for methane production)[d]	
			0.056 (combustion of natural gas for methanol production)			0.056 (combustion of natural gas for electricity production and air separation)	
Olefin Process	0.053 (combustion of fuel-grade byproducts)[e]	0.047 (combustion of fuel-grade byproducts)[f]	0.056 (combustion of natural gas and a small amount of fuel gas byproducts for product recovery and separation. However, for reactions, little combustion of fuels is needed since the exothermic reactions are largely self-driven)			0.012-0.018 for electricity production efficiency of 55-60% respectively (combustion of all non-methane and non-ethylene byproducts)[g]	0.050-0.054 for electricity production efficiency of 55-60% respectively (combustion of all non-ethylene byproducts and methane)[h]

[a] Data is based on (Joosten 2001). Refinery gases have higher carbon content than natural gas.
[b] Data is based on (Joosten 2001).
[c] Data is based on (Joosten 2001). Here we assume natural gas as the combusted fuel, which is typical for natural gas-fired heaters used in the chemical industry (GEMIS 2001).
[d] Data is based on (Joosten 2001).
[e] Data is based on (Matthews 2002).
[f] Data is based on (Matthews 2002). The difference of CO_2 emission factors between naphtha and ethane cracking is caused by the different fuels used. Ethane cracking emits 6 kg CO_2/GJ less than naphtha cracking does because the hydrogen content in the fuel grade byproducts from ethane cracking is higher than that from naphtha cracking (Matthews 2002).
[g] Data is based on (Swanenberg 1998). This is the result of partial oxidative, exothermic reactions in which thermal energy is released without producing CO_2 (See Table 2).
[h] Data is based on (Swanenberg 1998). The emission factor of OCM II is higher than OCM I because of less recycling.

provide heat needed) (Lunsford 2000). Steam reforming accounts for 50-70% of the total exergy losses (due to combustion of methane) in the current methanol production (Machida et al. 1998; Hugill et al. 2001).

There are a few alternative processes to the methanol routes although all of them are in early stages. Syngas can be directly converted to olefins, but low selectivity is still a problem[1]. Another possibility is to skip methanol as intermediate and produce DME directly from natural gas with less energy use than indirectly via methanol (Liu et al. 2000). However, we did not find yet energy use data. Tandem reforming (patented by Russia's GIAP and licensed by Linde AG (Michel 1999)) is reported to use approximately 20% less energy than traditional steam reforming and could result an overall energy use of approximately 7-8 GJ/t methanol. Another so-called "sono-chemical" process converts methane to methanol using high-intensity ultrasound (Casadonte et al. 2001). However, we did not find energy use data.

Second, dehydration catalysts used in methanol routes lead to additional energy use and CO_2 emissions that steam cracking does not. Metal oxide catalysts generally have problems like erosion, coking[2], temperature shocks, attrition (due to suboptimal residence time) and disposal problems (due to heavy metal contents). The efficient use of catalysts requires decoking and regeneration (usually at a high temperature range of 450-700 °C) (Chen et al. 1998; USPTO 2002). In the case of steam cracking, catalyst is only used in acetylene separation.

Third, as an advantage, methanol routes have much less burden in terms of product separation, recovery, transportation, distribution and effluent treatment than steam cracking does since there are fewer byproducts from methanol routes than from steam cracking (see Table 2). Also, the level of acetylenic and diolefinic compounds in the effluent from MTO units are significantly lower compared to steam cracking, resulting additional simplification in the recovery section (Kvisle et al. 2001). It is reported that emissions of nitrogen and sulfur oxides, particulates (causing acid rain and pollution) and greenhouse gases from the natural gas-to-olefin/gasoline chain are about 25% to 99% lower than the emissions from the crude oil-to-liquid fuels chain (EIA 2003).

In a word, methanol routes to olefins are burdened with relatively high energy use and CO_2 emissions (due to the use of methanol and catalysts) although there are some other environmental advantages in comparison with steam cracking. Therefore, improvement in energy efficiency should come from further R&D. Given the 40-year long R&D on steam cracking as opposed to that of 12-year on methanol-to-olefins, this is hopeful.

---

[1] Syngas to olefin routes can be direct (Stratton 1983; USPTO 1986). Direct synthesis of syngas to olefins (a variation of Fischer-Tropsch technologies) is difficult to control with current catalysis and often results in polymerization with low selectivity to light olefin and high yield of heavy hydrocarbons. This problem is known as the rule of Anderson-Schulz-Flory distribution function (ASF) (Wang et al. 2003). Therefore, it has been around for sixty years, but there is yet no sign of near commercialization for olefin production (Keep 1999).

[2] Lurgi claims that their MTP process employs a fixed bed reactor and might have fewer coking problems than UOP MTO's fluidized reactor does(Holtmann et al. 2001). Experts from UOP LLC argue that UOP MTO has coke yield 2-3%, which is lower than that of Lurgi MTP (Vora 2004).

5.2 OCM Routes

The current performance of OCM routes in terms of energy use is poorer than state-of-the-art steam cracking primarily due to the following problems:

- **Low yields** There is a trade-off between methane conversion and selectivity to ethylene (Choudhary et al. 2004). The rate of reaction is slow under 600 °C, but above 600 °C undesired oxidations tend to dominate (Weissermel et al. 2003);
- **Separation** Relatively high energy use in separation and resulting CO_2 emissions (Gonzalez 1999; Choudhary et al. 2004);
- **Catalysts** Additional oxygen and hydrogen are required for reducing coking on catalysts (Maitra 1993; Lunsford 1995); High temperatures (approximately 750-1,000 °C) require catalysts with high thermal stability (Choudhary et al. 2004);

Other issues concern safety (possible explosion due to the mixture of oxygen and hydrocarbons), environmental pollution (CO/CO_2 emissions, aromatic hydrocarbons, acid gas and organic acid), and operability (Zaman 1999).

Currently, a number of methods are being experimented to tackle these problems:

- **Raising Yield** New energy-efficient catalysts that can raise conversion and selectivity above 20-30% and 80-90% respectively (Choudhary et al. 2004; Pyatnitskii 2003); Oxygen-permeable membrane reactors are being researched to eliminate the need for oxygen plants and to enhance the selectivity to ethylene (Lu 2000). Up to 50-60% ethylene yield is possible (Lu et al. 1997). However, the slow oxygen flux will require a large reactor (Hugill 2003). If oxygen-permeable membrane reactors could be integrated with an OCM plant, they would allow approximately 20% (approximately 6.5 GJ/t ethylene) energy savings compared to a DSM OCM I plant (without membrane reactors) (Lu et al. 1997; Hugill 2003); Hydrogen (much more than produced in OCM required) can be used to converted carbon oxides (byproducts) into methane (so-called carbon conservation) (USPTO 2000). It is also possible to use hot hydrogen to convert carbon oxides to gasoline and methanol (USPTO 1991);
- **Membrane separation** Another membrane application is in recycling and separation sections where ethylene is removed from end product streams. Ethylene yield up to 75% has been claimed (Cordi et al. 1997).
- **Retrofitting** In addition, OCM can alternatively be retrofitted with conventional ethane or naphtha steam crackers for energy optimization (USPTO 1997; USPTO 1993; Hoebink et al. 1995; Czechowicz et al. 2004).

If implemented, these new methods could bring large improvement in terms of yield, energy use and CO_2 performance. Yet at this moment, none of these possibilities have been demonstrated and commercialized. However, partial oxidation is highly exothermic (releasing thermal energy without producing CO_2) and could become a low-CO_2 electricity and chemical co-generating technology. This will encourage future research on the process.

6 Conclusion

Recent development of C_1 technologies has motivated our energy and emission analysis in comparison with steam cracking. C_1 technologies are interesting because methane as an alternative feedstock is utilized and other sources such as biomass, coal and unconventional gas reserves could serve as additional source for methane. However, our energy analysis shows that even the most efficient natural gas-to-olefins-via-methanol uses more than twice as much primary energy as state-of-the-art steam cracking to produce one ton of high value chemicals. This is largely due to high energy use in methanol production. There is some potential in efficiency improvement, but no breakthrough is yet in sight. Two OCM designs are studied and it is found that CO_2 emissions can be significantly lowered if more recycling is used and less electricity is co-generated.

Appendix I: Definitions of Energy Uses in Steam Cracking

In this paper, various processes are compared with regard to their *cumulative energy use*. The *cumulative energy use* of a given system is the sum of all *net energy use* for all processes in the system (process chain analysis). *Net energy use* for each process can be determined through the making of an energy balance, i.e. subtracting the total energy output from the total energy input and relating the outcome to the physical production. This approach is commonly used in life-cycle assessment (LCA) and energy analysis. In most cases, this *net energy use* is identical with *process energy use*, which represents the total of directly consumed fuels, electricity and steam (in final or primary terms). This applies to all processes discussed in this paper except for steam cracking. The reason for this difference and consequences for the results are discussed in this appendix.

Steam cracking is an exception because its *net energy use* (as defined above) is lower than its *process energy use*. Steam cracking is a highly endothermic process. Figure 5 shows that the endothermicity of naphtha cracking is approximately 8 GJ/t ethylene (in the case of ethane steam cracking, this figure is approximately 5 GJ/t ethylene). Endothermicity is a minimum thermodynamic theoretical energy required to convert naphtha or ethane to desired chemicals. This energy (endothermicity) is not lost, but ends up in the final products (which have a higher specific calorific value expressed in GJ/t final products).

When calculating *net energy use* in the case of steam cracking, we subtract the total energy output from the total energy input in Figure 5: 150 – 128 = 22 GJ/t ethylene. In contrast, *process energy use* in naphtha and ethane-based steam cracking is the sum of *net energy use* and endothermicity (see Figure 5: fuel = 30 GJ/t = 8 + 22) (For steam cracking, the *process energy use* is therefore larger than the *net energy use*).

The *process energy use* for naphtha cracking shown in Figure 5 represents the world average (30 GJ/t ethylene). The respective value for state-of-the-art naphtha cracking is 20 GJ/t ethylene. Expressed in terms of high value chemicals (100% weight given to aromatics), the cumulative energy uses are respectively 23 GJ/t HVC and 16 GJ/t HVC. This is 5 GJ/t HVC higher than using the *net energy use* of naphtha-based steam cracking shown in Table 3.

Similarly, in the case of ethane cracking, the *process energy use* for world average ethane cracking is 20 GJ/t ethylene and 15 GJ/t ethylene for ethane-based state-of-the-art steam cracking. The cumulative energy uses are respectively 23 GJ/t HVC and 17 GJ/t HVC. The *process energy uses* are hence 5 GJ/t HVC higher than using the *net energy use* shown in Table 3.

In steam crackers, *process energy use* can be easily monitored and is therefore generally used to measure energy efficiency. Moreover, *process energy use* is directly related to CO_2 emissions and it determines most of production costs. For detailed discussion and calculations regarding this issue, please see (Ren et al. 2004).

Figure 6 shows a comparison of the cumulative energy use (based on the definition of *process energy use*). In contrast, Figure 3 shows a comparison of the cumulative *energy use* (based on the definition of *net energy use*). According to Figure 3, state-of-the-art naphtha and ethane steam cracking are the most advantageous. We consider that Figure 3 represents a more valid finding than Figure 6 because in the former a comprehensive energy balance is used.

Figure 5 Energy Flow (GJ/t ethylene) in a World Average Naphtha-based Steam Cracker

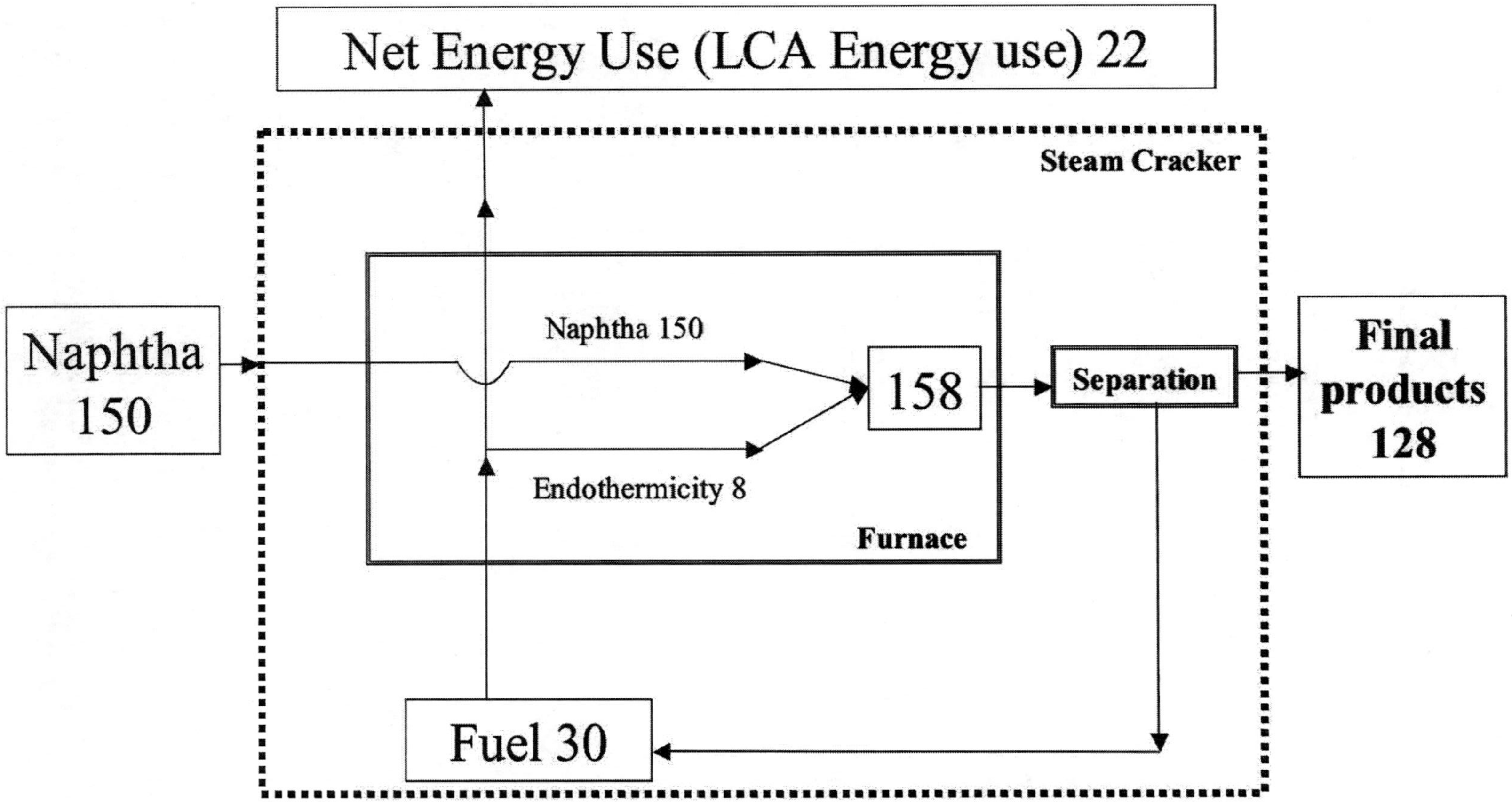

Figure 6 Comparison of Cumulative Energy in Steam Cracking and Natural Gas-to-Olefins (error bars represent the same sensitivity analysis as Figure 3 and 4)

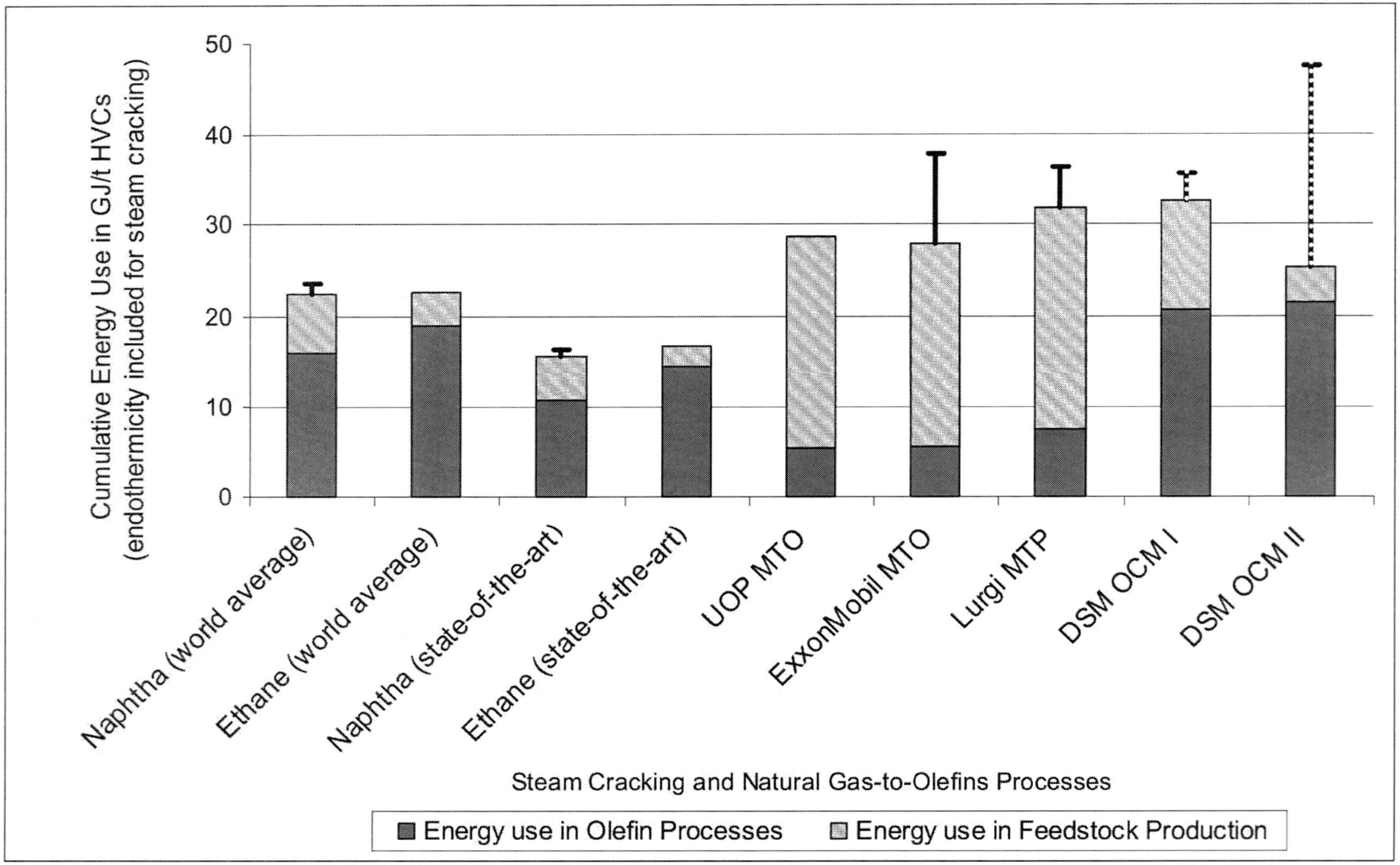

References

Addroulakis, I. and S. Reyes. Optimal design and operation of methane oxidative coupling reaction systems for the production of higher hydrocarbons. Technical Program at the Annual Meeting of the American Institute of Chemical Engineers (AIChE), Houston, American Institute of Chemical Engineers. 2000.

Albright, L. F., B. L. Crynes and S. Nowak, Eds. Novel production methods for ethylene, light hydrocarbons and aromatics. New York, Marcel Dekker. 1992.

Alperowicz, N. Egyptian Firm to Build MTO Complex. Chemical Week. See also: News report on April 9, 2003 at www.chemweek.com. 2003.

Alperowicz, N. Methanol-to-Olefins Complex Planned for Nigeria. Chemical Week. See also: News report on October 2, 2002 at www.chemweek.com. 2002.

Barger, P. UOP Presentation. The IZA Pre-conference School on July 5-7, Poitiers, France. 2001.

Bonnekamp, P. and H. Mannaert. Protocol Monitoring Energiebesparing, CPB/ECN/Novem/RIVM (ECN-C-120, RIVM 408137005). 2001.

BUWAL. Bundesamt fur Umwelt, Wald und Landschaft (in German). Bern, Vergeleichende okologische Bewertung von Anstrichstoffen im Baubereich. Band 2. Daten Scheriftenreihe Umwelt Nr. 232. 1995.

Casadonte, D. and M. Donahue. Oxidative Sonochemical Production of Methanol from Methane (file TTU D-0164), Office of technology transfer and Intellectual property, Texas Tech University System. 2001.

Chang, C. Mechanism of hydrocarbon formation from methanol. Methane Conversion. D. Bibby, C. Change and S. Yurchak. Amsterdam, Elsevier Science Publishers: 127-143. 1988.

Chauvel, A. and G. Lefebvre. Petrochemical processes, Part I: Synthesis gas derivatives and major hydrocarbons. Paris, Editions Technip. 1989.

ChemSystems. 10-year Proprietary Report Summaries. Houston, ChemSystems Inc. 2003.

ChemSystems. Process Evaluation/Research Planning: PERP 2002/2003 Program (Appendix II Production Cost Tables). Houston, ChemSystem/Nexant Inc. 2002.

Chen, D. and H. Rebo. DME conversion to light olefins over SAPO-34: deactivation due to coke deposition. Natural Gas Conversion. A. Parmaliana. Amsterdam, Elsevier Science BV. 1998.

Choudhary, V. R. and B. S. Uphade. "Oxidative conversion of methane/natural gas into higher hydrocarbons." Catalysis Surveys from Asia 8(1): 15-25. 2004.

CMAI. Europe/Middle East Report (olefins & derivatives) Issue No. 200, Chemical Market Association Inc. 2003.

CMAI. European Market Report (aromatics) Issue No. 181, Chemical Market Association Inc. 2003.

Cordi, E., S. Pak, M. Rosynek and H. Lunsford. "Steady-state production of olefins in height yields during the oxidative coupling of methane: utilization of a membrane contractor." Applied Catalysis 155: L1-L7. 1997.

Costanza, R. and R. Herendeen. "Embodied energy and economic value in the US economy: 1963, 1967 and 1972." Resources and Energy 6: 129-163. 1984.

Czechowicz, D., K. Skutil, A. Torz and M. Taniewski. "An Integrated process of oxidative coupling of methane and pyrolysis of naphtha in a scaled-up unit." Journal of Chemical Technology and Biotechnology 70: 182-186. 2004.

Dermaut, J. and B. Geeraert. A better understanding of greenhouse gas emissions for different energy vectors and applications. The 17th WEC Congress, Houston, The World Energy Council. 1998.

DOE. Manufacturing Consumption Energy Survey, Energy Information Administration of US Department of Energy. Also available at www.eia.doe.gov/emeu/mecs/iab98/chemicals/index.html. 1998.

Dubois, D. R., D. L. Obrzut and J. Liu. "Conversion of methanol to olefins over cobalt-, manganese- and nickel-incorporated SAPO-34 molecular sieves." Fuel Processing Technology(83): 203-218. 2003.

EC-BREF. Integrated pollution prevention and control (IPPC) Reference Document on Best Available Techniques in the Large Volume Organic Chemical Industry February 2002, Lower Olefins Process BREF. Seville, Institute for Prospective Technological Studies (Technologies for sustainable development) European IPPC Bureau. 2002.

EIA. International Natural Gas Information, Energy Information Administration at the US Department of Energy. 2003.

Energetics. Energy and Environmental Profile of the U.S. Chemical Industry. Columbia, Maryland, Study prepared by Energetics Co. for the Office of Industrial Technology at the US Department of Energy. 2000.

EPO. ICI Olefins (European Patent Application 40,015). European Patent Office, ICI. 1981.

Geem, P. "Nieuwe routes naar etheen." NPT ProcesTechnolgie (in Dutch). A journal jointly published by the KIVI Afdeling voor Procestechnologie, de KNCV sectie Procestechnologie en de NIRIA Vaksektie Procestechnologie(April): 15-22. 1992.

Geerts, J. Ethylene from natural gas by direct oxidation (Report: EUR 1361 EN). Eindhoven, The Netherlands, Eindhoven University of Technology. 1991.

Geerts, J., J. Hoebink and K. Wiele. Ethylene from natural gas: proven and new technologies. Novel Production Methods for Ethylene, Light Hydrocarbons and Aromatics. b. C. a. s. N. L.Alibright. New York, Marcel Dekker: 207-237. 1992.

GEMIS. Global Emission Model for Integrated Systems. Darmstadt, Germany, Institute for Applied Ecology (Oko-Institut). 2001.

Gonzalez, L. Conversion of natural gas into heavier hydrocarbons: a process scheme. The 78th Annual Convention of Gas Processors Association, Nashville, Tennessee, Gas Processors Association. 1999.

Gradassi, M. and N. Green. "Economics of natural gas conversion processes." Fuel Processing Technology 42: 65-83. 1995.

Gregor, J. Meeting the changing needs of the olefins market by UOP LLC. 5th EMEA Petrochemical Technology Conference, June 23-26, Paris, Euro Petroleum Consultancy Ltd. 2003.

Herz, R. Methanol Conversion 2000, Chemical Process Laboratory, University of California in San Diego. 2000.

Hoebink, H., H. Venderbosch and P. Geem. "Economics of the oxidative coupling of methane as an add-on for naphtha cracking." Chemical Engineering and Technology 18(1): 12-16. 1995.

Holtmann, H. D. and M. Rothaemel. Lurgi's Route from methanol to propylene. ERTC Petrochemical Conference on Feb. 21-23, 2001, Rome, Italy, Global Technology Forum. 2001.

Huff, G. A. and I. A. Vasalos. "Oxidative pyrolysis of natural gas in a spouted bed reactor: reaction stoichiometry and experimental reactor design." Catalysis Today 46(1998): 223-231s. 1998.

Hugill, J. Personal communications in various occasions between 2002 and 2004 regarding an ECN study on Swanenberg's work (1998) on oxidative coupling. Petten, The Netherlands, ECN (Energy Research Foundation of the Netherlands). 2003.

Hugill, J., J. Overbeek and S. Spoelstra. A comparison of the eco-efficiency of two production routes for methanol (ECN-I-01-003). Petten, The Netherlands, ECN (Energy Research Center of the Netherlands). 2001.

Hydrocarbon-Processing. Petrochemical Processes 2003. Hydrocarbon Processing. 2003.

Hydrocarbon-Processing. Refining Processes 2002. Hydrocarbon Processing. 2002.

IEA. Energy Statistics of OECD countries 2001/2002 and Energy Statistics of Non-OECD Countries 2001/2002 (2004 Edition), International Energy Agency, Paris. 2004.

IFE. The Gas Value Chain: technology strategy for competitive gas production and off-take, Norwegian Institute for Energy Technology. See also: http://www.ife.no/media/1234_OG_21-2003-Gassverdikjeden.pdf. 2003.

IKARUS. Personal communication with Linde AG. Quoted in IKARUS (Instrument for climate changing gas reduction strategies) (in German), FhG-ISI, GEU, FfE and Oko Institute. 1994.

IPCC. The Third Assessment Report: Climate change 2001--Mitigation. Geneva, Switzerland, Intergovernmental Panel on Climate Change. 2001.

Joosten, L. The Industrial Metabolism of Plastics; Analysis of Material Flows, Energy Consumption and CO2 Emissions in the Lifecycle of Plastics. P. d. W. C. Turkenburg and D. E. Worrell. Utrecht, The Netherlands, Dept. of Science, Technology and Society, Utrecht University. 2001.

Keep, P. Comparison of remote gas conversion technologies, Synetix Inc. 1999.

Keller, G. E. and M. M. Bhasin. "Synthesis of Ethylene via Oxidative Coupling of Methane." Journal of Catalysis(73): 9-19. 1982.

Koempel, H., W. Liebne and M. Wagner. MTP--An economic route to dedicated propylene. The 2nd ICIS-LOR World Olefins Conference, Amsterdam, ICIS-LOR Inc. 2003.

Kunii, D. and O. Levenspiel. Free Fall Reactor. New York, Robert E. Krieger Publishing Co. 1977.

Kvisle, S. and H. R. Nilsen. MTO: state of art and perspectives. DGMK-Conference: creating value from light olefins-production and conversion on October 10-12, Hamburg, Germany, German Society for Petroleum and Coal Science and Technology. 2001.

Liebner, W. GTC- Gas to Chemicals Process Options for Venezuela by Lurgi Oel-Gas Chemie Engineering. PdVSA-EFO Seminar, 6 March 2002, Caracas, Venezuela. 2002.

Liu, Z., C. Sun and G. Wang. "New progress in R&D of lower olefin synthesis." Fuel Processing Technology 62: 162-172. 2000.

Lu, Y. "Oxygen-permeable dense membrane reactor for the oxidative coupling of methane." Journal of Membrane Science 170: 27-34. 2000.

Lu, Y., A. G. Dixon, W. r. Moser and Y. H. Ma. "Analysis and optimization of cross-flow reactors for oxidative coupling of methane." Industrial Engineering Chemistry Research 36: 559-567. 1997.

Lunsford, J. H. "Catalytic conversion of methane to more useful chemicals and fuels: a challenge for the 21st century." Catalysis Today 63: 165-174. 2000.

Lunsford, J. H. "The Catalytic Oxidative Coupling of Methane." Angewandte Chemie International Edition 34: 970-980. 1995.

Lurgi. Lurgi MegaMethanol brochure by engineering division of Lurgi Oel, Gas and chemie GmbH. See also www.lurgi.de, Lurgi GmbH. 2003.

Machado, G., R. Schaeffer and E. Worrell. "Analysis: energy and carbon embodied in the international trade of Brazil: an input-output approach." Ecological Economics 39: 409-424. 2001.

Machida, S., J. Yagi and T. Akiyama. Exergy analysis of methanol production system (Kagaku Kogaku ronbunshu in Japanese). Tokyo, Japan, New Energy and Industrial Technology Development Organization (NEDO). 24. 1998.

Maitra, A. M. "Critical Performance Evaluation of Catalysts and Mechanistic Implications for Oxidative Coupling of Methane." Applied Catalysis 104: 11-59. 1993.

Matthews, R. Personal communications with Dr. Martin Patel from the STS department at the Utrecht University in the first half of 2002. London, England, Guthrie-Matthews Consulting Limited. 2002.

Michel, S. Methanol Production Costs, Reports on Science and Technology 61/1999, Linde. 1999.

Moiseev, I. I., N. A. Plate and G. F. Tereshchenko. "Energochemical Technologies: energy and bulk/basic chemicals in a single process." Russian Chemical Bulletin (International Edition) 50(11): 2015-1019. 2001.

Nakamura, D. "Special report: Ethylene capacity rising, margins continue to suffer." Oil & Gas Journal(March 11, 2002). 2002.

Neelis, M., M. Patel and P. Bach. Inventory of processes in the chemical and refinery industry (report Nr. NW&S-E-2003-61). The Netherlands, Dep. of Science, Technology and Society at the Utrecht University and Energy Research Center of the Netherlands. 2003.

Patel, M. "Cumulative energy demand and cumulative CO2 emissions for products of the organic chemical industry." Energy 28: 721-740. 2003.

Pujado, P. Personal communication through email between 2003 and 2004. Des Plaines, Illinois, UOP LLC. 2003.

Pujado, P., D. Greer, J. Andersen, T. Foley and V. Bhirud. The role of new technology in the light olefins industry. Spring National Meeting of the American Institute of Chemical Engineers (AIChE), American Institute of Chemical Engineers. 2002.

Pyatnitskii, Y. "Contemporary methods for the direct catalytic conversion of methane." Theoretical and Experimental Chemistry 39(4): 201-218. 2003.

Ren, T., M. Patel and K. Blok. Energy Use of Steam Cracking and Innovative Olefin Technologies. Energy Efficiency in Integrated Pollution Prevention and Control (IPPC) Installations on October 21-22, 2004, Vienna, Austria, Umweltbundesamt (Dep. of Energy in Austria). 2004.

Rogner, H. "An Assessment of World Hydrocarbon Resources." Annual Review of Energy and Environment 22: 217-262. 1997.

Rothaemel, M. and H. D. Holtmann. MTP, Methanol to Propylene--Lurgi's Way. DGMK Conference "Creating Value from Light Olefins--Production and Conversion", Hamburg, German Society for Petroleum and Coal Science and Technology. 2001.

Spath, P. and D. Dayton. Preliminary Screening-- Technical and Economic Assessment of Synthesis Gas to Fuels and Chemicals with Emphasis on the Potential for Biomass-Derived Syngas. Golden, Colorado, National renewable Energy Laboratory under the US Department of Energy. 2003.

Stratton, A., D. F. Hemming and M. Teper. Ethylene production from oil, gas and coal-derived feedstock. London, IEA Coal Research. 1983.

Swanenberg, G. M. J. M. Cogeneration of ethylene and electricity with oxidative methane coupling: study of technical feasibility and economic merit (A study sponsored by DSM, Sparqle International and SEP). Eindhoven, The Netherlands, Eindhoven University of Technology. 1998.

Synetix. Methanol plant technology, Synetix Inc. 2004.

Tabak, S. A. and S. Yurchak. "Conversion of methanol over ZSM-5 to fuels and chemicals." Catalysis Today 6(3): 307-327. 1990.

UNDP. World Energy Assessment - Overview. New York, United Nations Development Programme, United Nations Department of Economic and Social Affairs, World Energy Council: 28. 2000.

UOP. UOP/Hydro MTO Process Methanol to Olefins Conversion. Des Plaines, UOP LLC. 2003.

USPTO. Catalytic Production of olefins at high methanol partial pressures (US Patent application 20030109765). US Patent Office, ExxonMobil. 2003.

USPTO. Catalytic wall reactors and use of catalytic wall reactors for methane coupling and hydrocarbon cracking reactions (US patent 5,599,510). US Patent Office, Amoco Corporation. 1997.

USPTO. Conversion process by J. H. Brophy and R. P. Manning (US Patent 4,726,913). US Patent Office, BP. 1988.

USPTO. Methanol to Olefin process with increased selectivity to ethylene and propylene (US Patent 6,534,692). US Patent Office, UOP LLC. 2003.

USPTO. Methanol, olefin and hydrocarbon synthesis process (US Patent 6,444,712). US Patent Office, ExxonMobil Inc. 2002.

USPTO. Methods for converting light hydrocarbons to olefins, gasoline and methanol (US Patent 5,026,934). US Patent Office, Lyondell Petrochemical Company. 1991.

USPTO. Olefins process which combines hydrocarbon cracking with coupling methane (US Patent 5,254,781). US Patent Office, Amoco Corporation. 1993.

USPTO. Oxidative Coupling of methane with carbon conservation (US Patent 6,096,934). US Patent Office, UOP LLC. 2000.

USPTO. Process for converting natural gas to higher value products using a methanol refinery remote from the natural gas resource (US Patent application 20030045764). US Patent Office, ExxonMobil Chemical Company. 2003.

USPTO. Process for producing light olefins from crude methanol (US Patent 5,714,662). US Patent Office, UOP LLC. 1998.

USPTO. Process for producing olefins from natural gas (US Patent 5,025,108). US Patent Office, Institute Francais du Petrole. 1991.

USPTO. Process for the production of ethylene and other hydrocarbons from coal (US Patent 4,563,197). US Patent Office, US Department of Energy. 1986.

USPTO. Production of Light olefins (from methanol, etc.) (US Patent 4,499,327). US Patent Office, Union Carbide Co. 1985.

USPTO. Production of light olefins from oxygenate using framework gallium-containing medium pore molecular sieve (US Patent application 20030018231). US Patent Office, ExxonMobil Inc. 2003.

USPTO. Two-stage quench tower for use with oxygenate conversion process (US Patent 6,459,009). US Patent Office, UOP LLC. 2002.

USTDA. EATCO-Suez Petrochemicals Complex Gas to Polyolefin Project Final Report (Executive Summary). Washington DC, US Trade and Development Agency. 2000.

Vora, B. Personal communication on August 11, 2004 through emails. Des Plaines, Illinois, UOP LLC. 2004.

Wang, C., L. Xu and Q. Wang. "Review of directly producing light olefins via co hydrogenation." Journal of natural gas chemistry(12): 10-16. 2003.

Weissermel, K. and H. J. Arpe. Industrial Organic Chemistry, Wiley-vch GmbH & Co. KGaA. 2003.

Worrell, E., D. Phylipsen, D. Einstein and N. Martin. Energy use and energy intensity of the US chemical industry. Berkeley, Environmental Technology Division of Energy Analysis Department at the Lawrence Berkeley National Laboratory. 2000.

Zaman, J. "Oxidative processes in natural gas conversion." Fuel Processing Technology 58: 61-81. 1999.

Zenz, F. A. and D. F. Othmo. Fluidization and Fluid Partial Systems. New York, Reinhold Publishing C. 1960.

125a Simplified Heat Transfer Equations for Rotary Kilns

Louis E Herrington

A technical Paper presented by the Author at the AIChE 2004 Spring Meeting contained equations describing heat transfer in Pebble Beds. There was mention that the equations could be used to characterize heat transfer in Rotary Kilns. In this Paper the equations will be used in two applications for Rotary Kilns. The first application will be for heating a gas while cooling solids that pass through the Kiln. In a Publication by Kamal Adham, the Kiln design is completed using classic heat transfer methods. Data in Dr. Adham's publication is applied to a design using the new equations. For the second application, hot air will treat kaolin clay in a Rotary Kiln. Clay properties taken from the literature allow a prediction of the temperature profile along the Kiln. Film Heat Transfer coefficients taken from other publications are used in these applications. For both cases the impact of variations in gas flow will be described. The effect of particle size and shape on heat transfer is discussed.

Effective Plant-Wide Energy Management and Optimization

Annelize Victor
Aspen Technology, Inc.
Houston, Texas

Abstract

The recent rise in energy prices and general acceptance that higher prices are here to stay, coupled with increasing emissions constraints, are causing the process industries to take another look at energy management. In fact, over the past few months, many industry leaders have even been speaking publicly about their experiences in addressing these new energy challenges. One such venue was AspenWorld 2004, held in October 2004.

This paper provides an assessment of the current energy situation and, through examples taken from some of these company presentations, shows how AspenTech is helping companies become better equipped to manage these new challenges.

Introduction to the Energy Scene

For most of us, it isn't the absolute cost of a problem that causes us to sit up and take notice – it's more a sudden change in the cost. It comes as no surprise, then, that the recent significant rise in energy prices, coupled with other factors such as growing concerns over emissions, has led to a renewed interest in energy efficiency and energy management.

The process industries depend on an adequate and reliable supply of energy in the form of fuel, steam and electricity to drive the production processes. For many in this industry, the operating cost associated with supplying this energy is second only to the purchase of raw materials as the highest cost item. In the oil refining industry, the cost of energy can exceed 60% of the total operating cost (excluding crude oil purchases), while in the chemicals industry, a corresponding figure of 40% is not uncommon.

During the first three quarters of the twentieth century, the price of crude oil remained low and stable at under $5 per barrel. Then, in the mid to late 1970s, the price of oil escalated dramatically, rising to over $30 per barrel by the early 1980s. This sharp rise in energy prices instigated a flurry of activity as companies tried desperately to retrofit plants designed on the basis of oil at $5 per barrel to the new world economics. By the mid 1980s, the price had fallen to a more respectable $15 per barrel – and despite many external influences, they hovered around this value until the end of the century.

The new century saw instability returning to the oil markets and oil prices rose to heights last seen in the early 1980s. Seeing this rise in energy prices, many companies were faced with a potential dilemma: should we invest now to improve energy efficiency, or should we ride the storm and hope that the days of plentiful supply of low price oil will return?

The Current Situation: What is Different Now?

In deciding the appropriate action (or indeed, *inaction*) we need to look at the current situation, compare it to the previous "energy crises," ask what – if anything – is different today, and in so doing attempt to predict the future.

Is the High Price of Energy a Concern within Our Industry?

A recent survey undertaken by Aspen Technology among many of our clients confirmed that energy cost is consistently within the top three or four key factors impacting company performance. Rick Fontenot, Vice President of Lyondell Equistar, summarizes the sentiments of many chemical companies today when he says, *"Record high costs for feedstocks and energy, future projections at continued high levels, and dramatic price fluctuations are real concerns for the chemical industry."*

Will the Recent and Dramatic Rise in Energy Costs be Followed by a Dramatic Fall to Previous Levels?

While it is impossible to predict the future with any degree of certainty, there is growing acceptance within the industry that the days of low energy prices are over. "*OPEC won't allow crude to fall below $40 a barrel*," said Phil Flynn, Vice President for risk management with the Chicago-based Alaron Trading Corporation in January 2005.

While the high oil price may itself be sufficient to instigate a program to improve energy efficiency, there are other factors that are contributing to making energy efficiency and energy management a "must have" rather than a "nice to have." These factors did not have the same prominence and impact last time around.

Environmental Concerns and Emissions Reduction

For some time, the process industries have had to deal with SO_x emissions, but increasingly NO_x and more recently CO_2 emissions are impacting the industry. Within the European Union, CO_2 trading has started and the energy-intensive industries have been allocated maximum site CO_2 emissions. A site in danger of exceeding its allocation will either have to buy CO_2 credits from the market, reduce throughput to curtail emissions or face a fine on excess CO_2 – currently set at €40/tonne of CO_2. Conversely, a site that is able to operate with less than its CO_2 allocation will be able to sell CO_2 credits into the market. CO_2 trading is therefore likely to become an everyday trading activity for many companies.

Similarly NO_x emissions are being curtailed worldwide, either by the imposition of strict emissions limits or by similar emissions trading schemes.

As the utility and energy systems are often the major source of SO_x, NO_x and CO_2 emissions, the control of these emissions and management of credits are inextricably linked with energy management.

Deregulation of the Utilities Markets

Not long ago, large, often state-owned, and highly-regulated utilities suppliers were the norm worldwide. The move to a deregulated utilities market is slowly but consistently advancing across the globe. Deregulation was supposed to deliver lower cost energy supplies to all, using "consumer choice" as the means to reduce costs. While, in general, deregulation is achieving this objective, there have been some noticeable and perhaps unexpected side effects.

When a market is first deregulated, it often goes through a period of instability while the key players jockey for position and establish their optimum mode of operation in the new marketplace. This period of instability is characterized by wildly fluctuating utility prices, and companies not shielded by long-term contracts can be hard hit. Likewise, those companies who can exploit the market can make huge gains.

As the market settles down, more utility suppliers may enter the market. The existing suppliers combat increased competition by become more creative in the way they sell utilities. This, in turn, results in increasingly complex contract conditions and a greater focus on the user being able to specify in advance (or nominate) their requirements.

Companies no longer buy from a single supplier, but may have several suppliers for the same utility, with each supplier having a different form of contract. Process companies become utility suppliers by exporting "excess" utility, typically electricity, into the market. In this case, there is a clear need to be able to accurately predict the cost of the utility and the amount that can be made available for export.

Utility companies provide the investment to build and operate cogeneration plants on an industrial host site and share in the benefits. This may provide much needed investment in new infrastructure but may impose tighter requirements on the host company to accurately predict its utility requirements.

The net result of deregulation in all cases is choice and flexibility; however, with these come the need to better understand and manage the true site energy requirements.

Energy Challenges Facing the Process industries

Wait a minute! Haven't we just defined the challenges – rising energy prices, increasingly complex contracts and emissions?

Unfortunately, these are not the only challenges that face many of the process companies today. Other pitfalls lie in wait.

Awareness

In most companies, most people are unaware of the scale of the problem and hence do not know how to deal with it. Put simply, what you don't measure, you cannot manage.

Identification of the Owner of the Problem

Who owns the energy problem? Many people may be involved – from the purchaser who negotiates the contracts, to the power plant manager who generates the steam and power, to the process engineer who specifies the reflux rate on a distillation column, to the process operator who turns the valve, to the energy manager who, in his part time role, is supposed to manage a multi-million dollar budget – often without any real authority.

Understanding of the Magnitude of the Problem...and the Opportunity

Does the organization understand the size and nature of the problem, and has this been translated into a clear opportunity?

Goals

What are the goals of the company? How have these been set, are they realistic, what is the progress against plan and are the goals shared by all? Is the organization willing and able to make the necessary changes to meet these goals?

Definition of Opportunities and Their Solutions

The first step on the road to sustainable improvement in energy management is knowing where we are now. Today's actual operation will be a compromise that keeps us clear of many constraints and provides a comfort level in operation at a certain cost. This can be seen in **Figure 1**. While such an operating philosophy may have been acceptable during the period of low energy prices, in today's economic environment it is not. In the seemingly never-ending strive to improve profitability; we are obliged to constantly pursue areas to reduce costs. This means operating ever closer to constraints – energy performance management.

Energy Performance Management

Energy performance management is the ability to know precisely at any time what the energy situation is, how this compares to where we could or should be, the cost of deviation and, most important, be able to identify savings opportunities and take appropriate action.

For most energy-using operations, there is a minimum amount of energy required to carry out the operation to meet certain product specifications. Use any less and the operation fails to complete successfully. Using more, however, does not necessarily bring any additional benefits. This means we can define a minimum energy requirement (or target) and measure actual performance versus target – the difference being the savings potential.

Utilities Planning

Every process site worldwide sets plans and schedules for production. The plan may not always be strictly followed, but no one would argue that this is not an important business process.

This methodology is, however, sadly lacking when it comes to planning and scheduling the operation of the utilities and energy systems. Rather than synchronize a production plan with a plan for energy production and distribution, the operation of the energy systems is more often than not reactive to the needs of the process units. Not having sufficient energy available at any given time is not an option therefore excess capacity need to be maintained to ensure constant reliability.

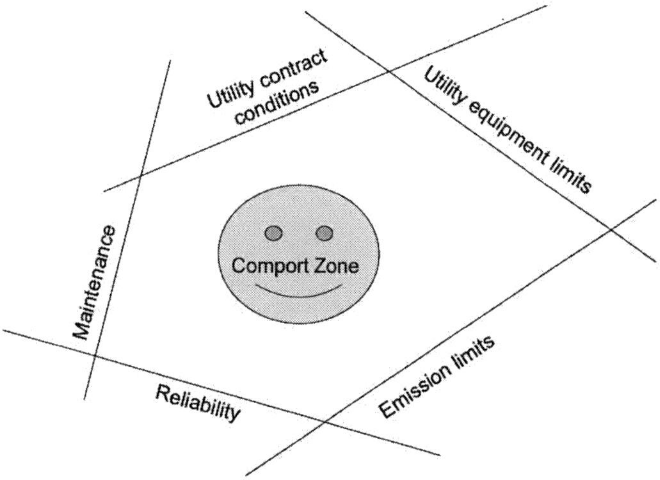

Figure 1: Actual Utility Plant Operation is Constrained by Many Factors

In this respect, excess capacity can be higher than necessary utility nominations to suppliers; boilers being on hot standby and steam turbines on slow roll. In all such cases, the provision of such "excess" capacity reduces energy efficiency and increases energy costs. Hence, the ability to forecast energy demands for the production units, coupled with knowledge of the supply options allows the setting of optimal, energy efficient utility production plans.

Optimal Operation

While setting a plan in advance is important, being able to execute against the plan is imperative if the benefits defined are to be achieved in practice. Optimal operation involves not only sharing of the plan with those in a position to implement but also the ability to modify the plan as conditions change.

Better Purchasing

Having an optimized energy production plan also benefits the procurement manager significantly. As previously stated, deregulation of the utilities markets worldwide has led to increased competition and the ability to purchase utilities from different suppliers with different contract terms. In general, the closer the company can define (or nominate) its import requirements, the cheaper the unit cost will become. Conversely, a contract providing highly flexible import capabilities will have a higher unit cost. It therefore follows that operation within the specified nomination range is of paramount importance. Therefore, the utilities operations manager's ability to accurately forecast demand and control the actual operation within the forecasted range will enable the procurement manager to buy cheaper than his peers who are not in such a fortunate position.

Investment in Energy Efficiency

Now we have the ability to optimize the daily operation of the energy infrastructure along with making optimized purchasing decisions. We have improved our baseline, how do become better yet? Should we consider investment to improve our inherent flexibility and therefore overall energy efficiency?

Investing in new energy efficient equipment can provide a step change in energy performance – but for every good investment in energy efficiency, there are at least twice as many projects that never achieve the predicted return on investment.

Escalating capital cost, oversizing of equipment and insufficient allowance for maintenance and down time can potentially reduce a good investment to one that is at best marginal. The ability to define the best investment opportunities and to accurately predict both total capital cost and net operating benefit are vital if the limited available capital is to be used wisely.

Energy Trading

For some companies, the ability to go one stage further than self-sufficiency, and actually export energy, can provide a lucrative additional revenue stream. In certain areas of the world it is the norm for oil refining and chemicals companies to export waste heat to adjacent companies or to the local community for space heating purposes. In Gothenburg, Sweden, the Volvo car factory and much of the city are supplied by waste heat from the local oil refineries.

As companies invest in cogeneration to supply their internal power needs, it can be tempting to make the cogeneration unit somewhat larger and provide an ability to export electricity. In all such export situations two pieces of information are vital (but often lacking). The first is how much energy can be exported and the second is the price cost of generating this energy or its internal value. Armed with this information the trader can decide when and how much energy to sell.

Continuous Improvement

It has often been stated that energy management is not a tactical task, but rather a never-ending journey. Unfortunately, while this is a true statement, it is also a philosophy that is rarely followed. In practice, energy efficiency and energy management are implemented as short-term objectives whenever there is a sudden rise in energy prices. This situation would not be too bad if it were not for one unfortunate factor – and that is, like any improvement initiative, the benefits of any energy management activity tend to reduce over time as people revert back to the previous way of working or changes to the plant are not accompanied by simultaneous changes to energy management systems and work processes. If the interval between energy management initiatives is sufficient the company sees no long term improvement but rather a saw-tooth effect. To prevent this happening, the systems and work processes need to be ingrained into all site personnel enabling it to achieve the state of unconscious competence in energy management.

An Energy Management Solution for Today

AspenTech has, over many years, developed a world-class capability in energy management and is today taking the lead in developing and implementing state-of-the-art energy management systems. Bringing together experience in energy management with accurate modelling and real time data acquisition, aspenONE™ Energy Management for Chemicals and aspenONE Energy Management for Petroleum provide a comprehensive step-by-step approach to delivering sustainable benefits.

Recognizing the site-specific yet common needs of industry, these solutions consist of a suite of capabilities that can be implemented individually or in any combination to address the specific site needs. The options of aspenONE Energy Management for Chemicals and aspenONE Energy Management for Petroleum are shown in **Figure 2** and the impact it has for our customers are summarized below.

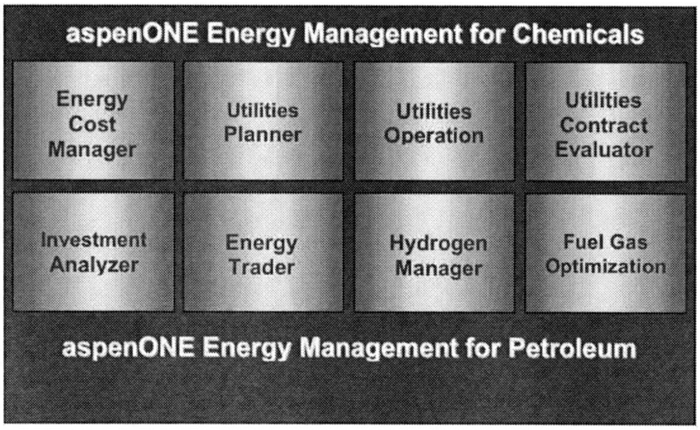

Figure 2: Components of aspenONE Energy Management for Chemicals aspenONE Energy Management for Petroleum

EniPower

EniPower is the owner and operator of the cogeneration plant on several refinery and petrochemicals sites in Italy. Deregulation of the electricity market in Italy started in 1999, with the Italian Electricity Deregulation Act signalling the end of the state monopoly, and culminated in the start up of the Market Pool on April

1, 2004. EniPower has declared its intention to be a significant player in the electricity export market, and by 2006 will have an installed electrical generation capacity of approximately 6,500MW, spread across seven sites – a fivefold increase on its installed capacity in 2002.[1]

In order to be able to trade effectively and profitably in the new deregulated market, EniPower defined the requirements of a Power Generation Management System which would provide EniPower with an integrated production information system to support the following business processes:

- Production planning
- Performance monitoring
- Production accounting and cost control
- Environmental monitoring.

The overall system, shown in **Figure 3**, provides updated data every 15 minutes on each site's performance and consolidates the information from each site back to headquarters in Milan. Web-based graphics are used to support visualization and information sharing.

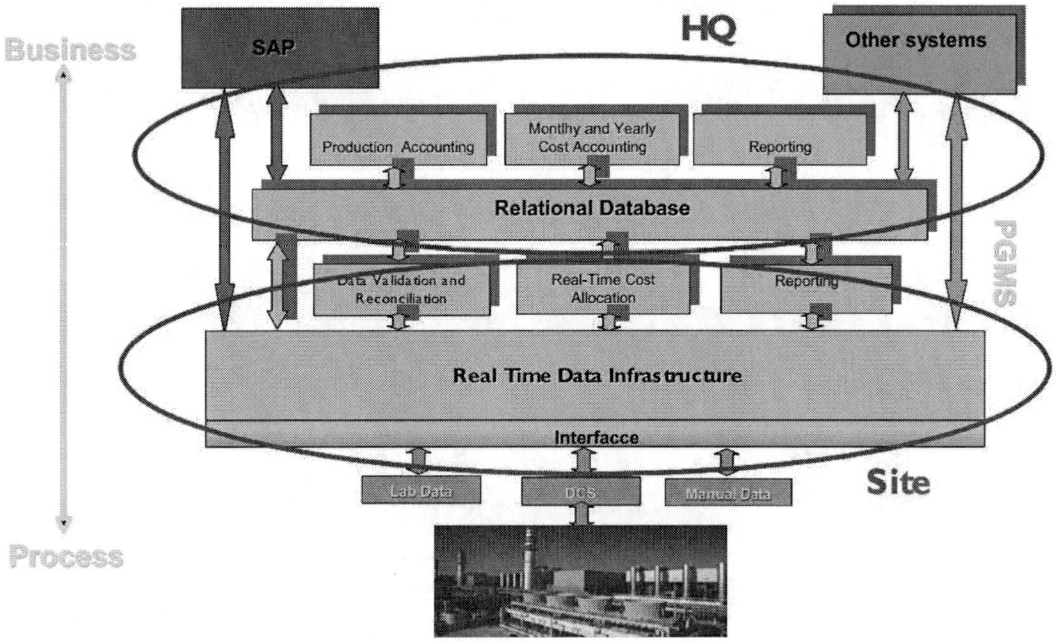

1

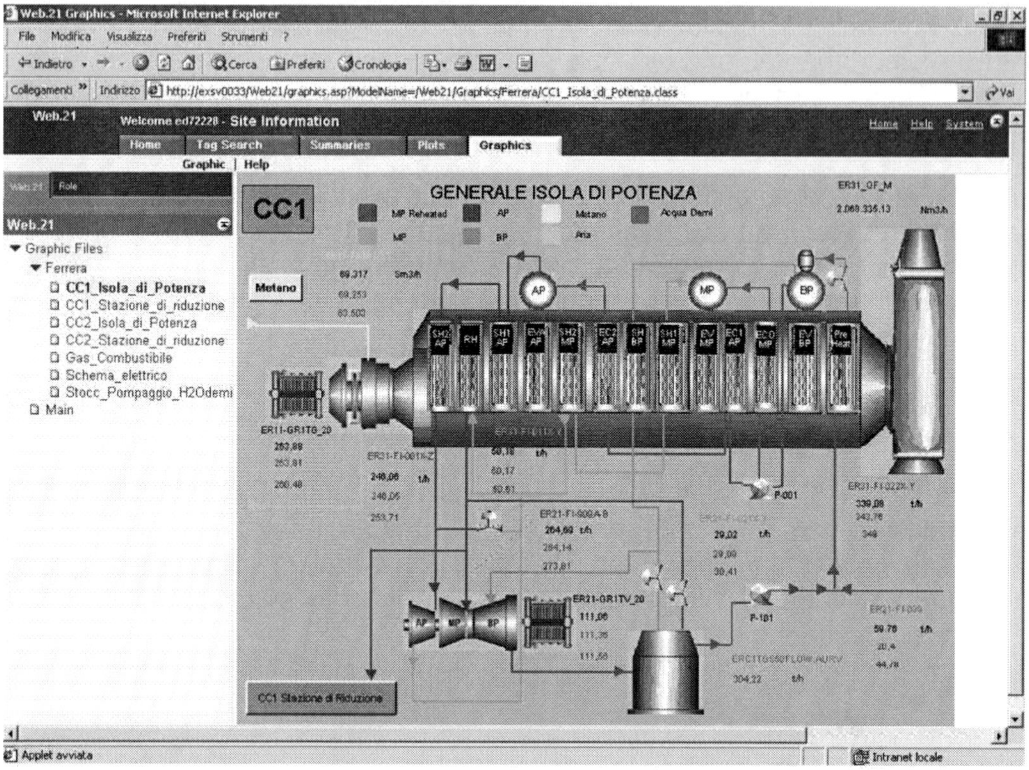

Figure 3: Architecture and Web-based Viewing at EniPower

Valero Energy Corporation

Valero Energy Corporation was founded in 1980 as a natural gas and oil refining company, and until 1997 owned just a single refinery. Over the past eight years, Valero has acquired 14 refineries and two asphalt plants, so that today it processes approximately 2.4 MMBPD of crude oil and feedstocks.

Valero has developed an energy strategy that will result in its becoming a first quartile performer in energy efficiency by 2009, and, in so doing, will also reduce greenhouse gas emissions by 4 MMT per year.

In 2002, Valero's Houston refinery embarked on a plan to improve its energy performance. An energy management system that would provide better planning and operation of the refinery energy systems was a key component of the plan.

The scope of the aspenONE Energy Management for Petroleum system implemented in the Houston refinery is shown in Figure 4, along with a typical comparison of the actual and optimized cost of operation – identifying significant operational improvements. Overall, Valero has identified potential first-year benefits at $2.7 million.

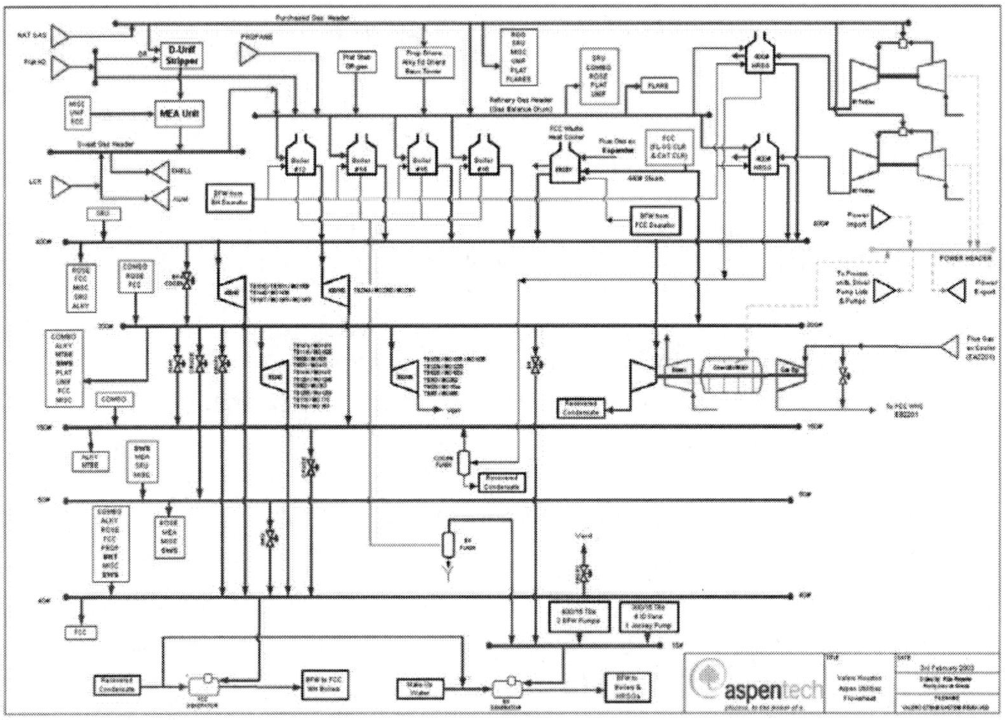

	Key Indicators of Utility System	Reconciled	Optimized	% Savings
Costs	Total Operating Cost	$ 2178 /hr	$ 1918 /hr	12%
	Cost of Natural Gas Import	$ 4084 /hr	$ 3925 /hr	4%
	Revenue from Sweet Gas Sales	$ 3225 /hr	$ 3225 /hr	0%
	Cost of Sweet Gas	$ 1188 /hr	$ 1188 /hr	0%
	Cost of (+) or Revenue from (-) Elecrticity	$ 131 /hr	$ 30 /hr	77%
Fuel	Total Natural Gas Consumption	17.7 MMscfd	17.0 MMscfd	4%
	Total Propane Consumption	0.0 BPD	0.0 BPD	
	Total Sweet Gas Used by site	28.6 MMscfd	28.6 MMscfd	0%
	Total Sweet Gas sales t	13.9 MMscfd	13.9 MMscfd	0%
Power	Power from Gas Turbine 1	12.5 MW	12.5 MW	0%
	Power from Gas Turbine 2	13.4 MW	13.4 MW	0%
	Power Import (+) or Export (-)	3.2 MW	0.7 MW	77%
	FCC Power Import (+) or Export (-)	4.3 MW	3.2 MW	25%
	FCC Expander Power	10.2 MW	10.6 MW	-5%
	FCC Turbo generator Power	0.7 MW	1.3 MW	-75%
	Switching Drive Power	4.3 MW	2.9 MW	33%
Steam	Steam Produced by HRSG 1	154.8 Mlb/hr	122.2 Mlb/hr	21%
	Steam Produced by HRSG 2	143.2 Mlb/hr	72.5 Mlb/hr	49%
	Steam Produced by Boiler 12	20.3 Mlb/hr	50.0 Mlb/hr	-147%
	Steam Produced by Boiler 14	20.7 Mlb/hr	50.0 Mlb/hr	-142%
	Steam Produced by Boiler 15	26.7 Mlb/hr	50.0 Mlb/hr	-87%
	Steam Produced by Boiler 16	25.0 Mlb/hr	50.0 Mlb/hr	-100%
	Steam Produced by FCC WHC	179.2 Mlb/hr	177.9 Mlb/hr	1%

Figure 4: Scope and Benefits at Valero's Houston Refinery

DSM

DSM is a multi-specialty chemicals company with 200 locations and 26,000 employees worldwide and sales of over €6 billion per year.

At its largest chemicals production facility in Geleen in southern Holland, DSM was experiencing similar issues to other leading chemicals companies; for example:

- Continual pressure for energy and cost reduction
- Increasing volatility in energy demand from the processes
- Increasing degrees of freedom
- Fuel selection
- Device selection (boilers, turbines etc)
- Market deregulation

In response to these issues, DSM defined the need for an integrated energy management system that would cover;

- Actual operational optimization to reduce daily costs
- Short-term tactical planning/optimization (1 - 30 days); e.g., in case of maintenance of boiler
- Longer term strategic planning (1 month - 5 years)
- Future contracts selection
- Evaluation of future investments

Benefits of the system have been quantified by DSM at 3 to 4% of total site energy cost with a return on investment of >25%.

DSM has been successful and combining operational, tactical and strategic requirements into a single Aspen Energy Management system as depicted in Figure 5.

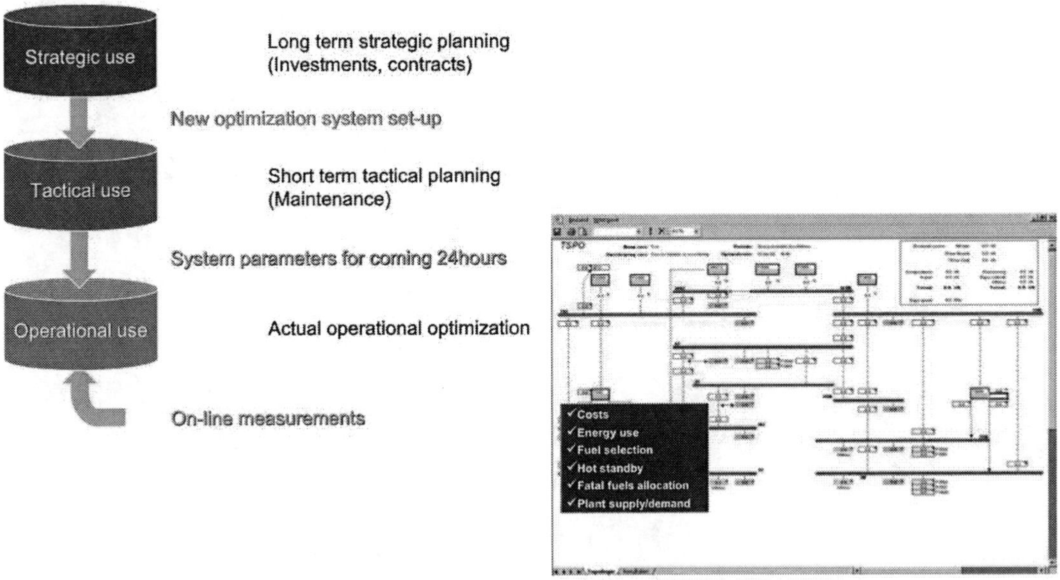

Figure 5: Multiple User Energy Management System at DSM

Conclusion

Rising energy prices are once again focusing companies in the process industries to look at how they source and use the energy that is vital to the processes. New issues such as strict emissions constraints, emissions trading and deregulation of the utilities supply markets are making energy management more complex that it has ever been before. The rewards of successful energy management can be great – but there are also many traps for the unwary.

AspenTech is helping industry by providing state-of-the-art tools that provide the much needed confidence and capability to tackle today's energy challenges.

References

Fontenot, Rick. AspenWorld 2004

Flynn, Phil. 2005

Lavecchia, Roberto. "Real Time Management of Large Cogenerative CCGT – The PGMS Experience." Orlando, Florida: Presentation at AspenWorld2004, 2004.

Faagau, Gary. "Can a Comples Refinery be Energy Efficient, Too?" Orlando, Florida: Presentation at AspenWorld 2004, 2004.

Schmitz, Jos. "Energy & Utilities Optimization: Experience with the AspenTech Energy Management and Optimization System." Orlando, Florida: Presentation at AspenWorld 2004, 2004.

125c Techno-Economic Evaluation of an Improved and Energy Efficient Natural Gas Liquid (NGL) Removal Process

Aqil Jamal, Dennis Leppin, Howard Meyer, Javad Abbasian, Yingzhong Lu

Natural Gas Liquids (NGL) is a collective term for C2+ hydrocarbons present in the natural gas. Historically, the commercial value of the separated NGL components has been greater than the thermal value of these liquids in the gas. The revenue derived from extracting NGLs is crucial to ensuring the overall profitability of the domestic natural gas production industry and therefore of ensuring a secure and reliable supply in the Lower-48. However, rising natural gas prices have dramatically reduced the economic incentive to extract NGLs from domestically produced natural gas. The successful gas processors will be those who will be using the technologies that are less energy intensive, have lower capital and operating costs and offer the flexibility to tailor the plant performance to maximize product revenue as market conditions change, while still maintaining efficient operation. The cryogenic turbo-expander technology is the dominant NGL recovery process throughout the world today. This process is known to be highly energy intensive, as it requires substantial energy to recompress the processed gas back to pipeline pressure. The purpose of this paper is to present the results of a techno-economic evaluation study based on computer simulation of a new NGL separation process that is flexible in terms of ethane rejection and reduces the current energy consumption by 15-20%, particularly at lower ethane recoveries (< 70%). The new process integrates the dehydration of the raw natural gas stream and the removal of NGLs in such a way that heat recovery is maximized and pressure losses are minimized and some high-value equipment such as the compressor, turbo-expander and a separate dehydration unit are altogether avoided.

125d Save Energy Without Capital Expenditure First before investing

Paul Tripathi

In these days of high enrgy prices and global comptition it is almost essential to save energy for the survival of the process industry. Eventhough significant energy savings could be achieved by capital projects getting the capital approved for the energy project is still very difficult. This paper will describe how you could gain significant energy savings withot any capital projects.

125e Improved Energy Efficiency Using Regenerative Oxidizers for VOC Control

Michael J Hager

Regenerative heat exchangers have largely replaced recuperative exchangers in thermal and catalytic combustion systems for VOC abatement. Typical regenerative designs provide thermal efficiencies in excess of 90% compared to 70% for plate type exchangers. MEGTEC Systems has developed a two-bed regenerative oxidizer that is capable of 99% VOC destruction by using a rotary valve to minimize the entrained volume. This compact design, however, presents a challenge in terms of achieving uniform flow distribution. CFD modeling and field tests have been conducted to optimize the media bed plenum design, and thus, maximize thermal efficiency. Several projects will be reviewed in which the installation of a MEGTEC Systems regenerative oxidizer has substantial reduced the customer's energy usage.

125f Energy Saving Opportunities in Heat-integrated Crude Oil Distillation Systems

Megan Jobson, Robin Smith, Vikas Rastogi

The crude oil distillation system in an oil refinery includes the atmospheric and vacuum distillation columns and the associated heat exchanger network. The distillation columns have significantly complex configurations and there are strong interactions between the distillation columns and the heat exchanger network. Changes to the distillation columns or the heat exchanger network can significantly influence the overall performance of the system. Therefore, it is important to analyse the distillation columns and the heat exchanger network simultaneously to explore opportunities for savings and profit enhancement.

Previous researchers developed a decomposition approach for simplified modelling of the atmospheric distillation column (Liebmann et el., 1998), and short-cut models for grassroots and retrofit design of the atmospheric distillation column (Suphanit, 1999; Gadalla et al., 2003a). Gadalla (2003b) also developed a simplified retrofit area-energy model for an existing heat exchanger network which correlates the additional area required in the heat exchanger network with energy consumption. This retrofit area-energy model does not consider the distribution of additional area in the heat exchanger network.

In the present work, the short-cut models for the atmospheric distillation column are extended to account for pressure drop and are developed further for simplified modelling of the vacuum distillation column. A simple model is developed for the vacuum steam ejector to calculate the steam requirements as a function of suction pressure at the top of vacuum column. The retrofit area-energy model for the existing heat exchanger network is converted to a retrofit cost-energy model, to incorporate the relationship between the distribution of additional area required in the heat exchanger network and capital cost.

The simplified models for the distillation columns and the existing heat exchanger network are applied with in a non-linear optimisation framework. The hydraulic limits of the distillation columns are imposed as optimisation constraints. The objective of the optimisation is to minimise the total annualised cost of the system, including the hot and cold utility costs, stripping and vacuum ejector steam cost, and the cost of additional area required in the heat exchanger network.

This paper illustrates the application of optimisation approach to an existing heat-integrated crude oil distillation system. The existing column configuration, i.e. existing number of stages and distribution of stages, pump-around locations, and side draw locations, is fixed. The feed preheat temperatures, stripping steam flow rates, reflux ratio, pump-around flow rates and temperature drops are optimised simultaneously in order to minimise the total annualised cost of the crude oil distillation system.

The case study shows that the operating cost can be reduced by 13% (2 MM$/yr) by changing the operating conditions of the distillation columns and making minor modifications to the heat exchanger network. In this case, a new heat exchanger is added and area is added to five existing heat exchangers.

The optimisation framework can be used to achieve other objectives, such as increasing profit, reducing greenhouse gases emission, throughput increase, etc.

References:

Liebmann, K., V. R. Dhole, and M. Jobson, Integrated design of a conventional crude oil distillation tower using pinch analysis, Chem. Eng. Res. Des., 76(A3), 335-347 (1998)

Suphanit, B., Design of Complex Distillation System, PhD Thesis, UMIST, Manchester, UK (1999)

Gadalla, M., M. Jobson, and R. Smith, Shortcut models for retrofit design of distillation columns, Chem. Eng. Res. Des., 81(A8), 971-986 (2003a)

Gadalla, M., M. Jobson, and R. Smith, Optimisation of existing heat-integrated refinery distillation systems, Chem. Eng. Res. Des., 81(A1), 147-152 (2003b)

125g Development of Advanced Membranes Technology Platform for Hydrocarbon Separations

Dilip Kalthod

The project objectives are to develop and demonstrate a new technology for superior, robust, low-cost membrane separator systems with multiple applications involving hydrocarbon separations, particularly natural gas dehydration, which has the broadest energy/ economic/ environmental impact in the U.S. and worldwide. Potential advantages of membranes over present technologies include lower methane and hydrocarbon loss per unit of natural gas production, lower energy consumption, reduced maintenance costs due to fewer components, no spent glycol/ molesieve/ deliquescent wastes, amenability to unmanned operation, smaller footprints, lighter weight packages, minimal electrical power duty, and amenability to expansion due to system modularity. The key task, critical to commercial success, is the development of a hollow fiber membrane with the necessary permeation properties and strength for operation in "dirty" streams. First an industry-based user-wants-and-needs study was conducted in order to provide and verify the technical and commercial targets for the new membrane technology. To help guide hollow fiber development, a computer model for membrane system performance was developed and used in optimization studies which included process and economic inputs such as system component costs, power consumption, methane loss. Spinning of several candidate fibers of the selected polymer at the pilot plant was started and the fibers were subjected to screening tests for permeation properties (H2O, air) and strength.

126h Ultra-Low Excess Air Control for Process Heaters

Kurt Jasin

Refinery Process Heaters are a large user of fuel. This proposal affords an opportunity to effect a major improvement in the combustion efficiency of process heaters and reduce fuel usage in the range of 1 to 5%. Side benefits are the possible reduction of prompt NOx of 50% from existing levels with or without low NOx burners. The proposal is to demonstrate the practicality of implementation of precise control for both stack flue gas dampers and inlet burner air registers to control near stoichiometric combustion. The control uses a fast light beam analyzer and a unique control strategy based on a trace amount of flue gas CO and highly sensitive furnace draft process variables to modulate the damper and air registers.

The technology can continuously operate heaters at 0.1 to 0.4% furnace excess O2 but must be demonstrated to be reliable and safe in the normal operating conditions of varying plant fuel gas Btu makeup and changes in heater firing rates. With proper demonstration of this technology in the industrial environment, the opportunity is for application across the Chemical and Petroleum industry as well as other industries with large fired heaters.